Mathematics and Logic

Mathematics and Logic

Mark Kac and Stanislaw M. Ulam

DOVER PUBLICATIONS, INC., NEW YORK

This Dover edition, first published in 1992, is an unabridged, unaltered republica-
tion of the work first published by Frederick A. Praeger, New York, 1968, under the
title *Mathematics and Logic: Retrospect and Prospects* and as "a *Britannica Perspective* prepared to commemorate the 200th anniversary of *Encyclopædia Britannica*."

Library of Congress Cataloging-in-Publication Data

Kac, Mark.
    Mathematics and logic / Mark Kac and Stanislaw M. Ulam.
        p.   cm.
    Originally published: New York : Praeger, 1968, in series: Britannica perspective.
    ISBN-13: 978-0-486-67085-0 (pbk.)
    ISBN-10: 0-486-67085-6 (pbk.)
    1. Mathematics—Popular works.   I. Ulam, Stanislaw M.   II. Title.
QA93.K24   1992
510—dc20   91-40429
                                                                         CIP

Manufactured in the United States by Courier Corporation
67085605
www.doverpublications.com

Introduction

WHAT IS MATHEMATICS? How was it created and who were and are the people creating and practising it? Can one describe its development and its role in the history of scientific thinking and can one predict its future? This book is an attempt to provide a few glimpses into the nature of such questions and the scope and the depth of the subject.

Mathematics is a self-contained microcosm, but it also has the potentiality of mirroring and modeling all the processes of thought and perhaps all of science. It has always had, and continues to an ever increasing degree to have, great usefulness. One could even go so far as to say that mathematics was necessary for man's conquest of nature and for the development of the human race through the shaping of its modes of thinking.

For as far back as we can reach into the record of man's curiosity and quest of understanding, we find mathematics cultivated, cherished, and taught for transmittal to new generations. It has been considered as the most definitive expression of rational thought about the external world and also as a monument to man's desire to probe the workings of his own mind. We shall not undertake to *define* mathematics, because to do so would be to circumscribe its domain. As the reader will see, mathematics can generalize any scheme, change it, and enlarge it. And yet, every time this is done, the result still forms only a part of mathematics. In fact, it is perhaps characteristic of the discipline that it develops through a constant self-examination with an ever increasing degree of consciousness of its own structure. The structure, however, changes continually and sometimes radically and fundamentally. In view of this, an attempt to define mathematics with any hope of completeness and finality is, in our opinion, doomed to failure.

We shall try to describe some of its development historically and to survey briefly high points and trenchant influences. Here and there attention will focus on the question of how much progress in mathematics depends on "invention" and to what extent it has the nature of "discovery." Put differently, we shall discuss whether the external physical world, which we perceive with our senses and observe and measure with our instruments, dictates the choice of axioms, definitions, and problems. Or are these in essence free creations of the human mind, perhaps influenced, or even determined, by its physiological structure?

Like other sciences, mathematics has been subject to great changes during the past fifty years. Not only has its subject matter vastly increased, not only has the emphasis on what were considered the central problems changed but the tone and the aims of mathematics to some extent have been transmuted. There is no doubt that many great triumphs of physics, astronomy, and other "exact" sciences arose in significant measure from mathematics. Having freely borrowed the tools mathematics helped to develop, the sister disciplines reciprocated by providing it with new problems and giving it new sources of inspiration.

Technology, too, may have a profound effect on mathematics; having made possible the development of high-speed computers, it has increased immeasurably the scope of experimentation in mathematics itself.

The very foundations of mathematics and of mathematical logic have undergone revolutionary changes in modern times. In Chapter 2 we shall try to explain the *nature* of these changes.

Throughout mathematical history specific themes constantly recur; their interplay and variations will be illustrated in many examples.

The most characteristic theme of mathematics is that of *infinity*. We shall devote much space to attempting to show how it is introduced, defined, and dealt with in various contexts.

Contrary to a widespread opinion among nonscientists, mathematics is not a closed and perfect edifice. Mathematics is a science; it is also an art. The criteria of judgment in mathematics are always aesthetic, at least in part. The mere truth of a proposition is not sufficient to establish it as a part of mathematics. One looks for "usefulness," for "interest," and also for "beauty." Beauty is subjective, and it may seem surprising that there is usually considerable agreement among mathematicians concerning aesthetic values.

In one respect mathematics is set apart from other sciences: it knows no obsolescence. A theorem once proved never loses this quality though it may become a simple case of a more general truth. The body of mathematical material grows without revisions, and the increase of knowledge is constant.

In view of the enormous diversity of its problems and of its modes of application, can one discern an order in mathematics? What gives mathematics its unquestioned unity, and what makes it autonomous?

To begin with, one must distinguish between its *objects* and its *method*.

The most primitive mathematical objects are positive integers 1, 2, 3, . . . Perhaps equally primitive are points and simple configurations (*e.g.,* straight lines, triangles). These are so deeply rooted in our most elementary experiences going back to childhood that for centuries they were taken for granted. Not until the end of the 19th century was an intricate logical examination of arithmetic (Peano, Frege, Russell) and of geometry (Hilbert) undertaken in earnest. But even while positive integers and points were accepted uncritically, the process (so characteristic of mathematics) of creating new objects and erecting new structures was going on.

From objects one goes on to sets of these objects, to functions, and to correspondences. (The idea of a correspondence or transformation comes from the still elementary tendency of people to identify *similar arrangements* and to abstract a common *pattern* from seemingly different situations.) And as the process of *iteration* continues, one goes on to classes of functions, to correspondences between functions (operators), then to classes of such correspondence, and so on at an ever accelerating pace, without end. In this way simple objects give rise to those of new and ever growing complexity.

The method consists mainly of the formalism of *proof* that hardly has changed since antiquity. The basic pattern still is to start with a small number of axioms (statements that are taken for granted) and then by strict logical rules to derive new statements. The properties of this process, its scope, and its limitations have been examined critically only in recent years. This study—metamathematics—is itself a part of mathematics. The *object* of this study may seem a rather special set of rules—namely, those of mathematical logic. But how all-embracing and powerful these turn out to be! To some extent then, mathematics feeds on itself. Yet there is no vicious circle, and as the triumphs of mathematical methods in physics, astronomy, and other natural sciences show, it is not sterile play. Perhaps this is so because the external world suggests large classes of *objects* of mathematical work, and the processes of generalization and selection of new structures are not entirely arbitrary. The "unreasonable effectiveness of mathematics" remains perhaps a philosophical mystery, but this has in no way affected its spectacular successes.

Mathematics has been defined as the science of drawing necessary conclusions. But which conclusions? A mere chain of syllogisms is not mathematics. Somehow we select statements that concisely embrace a large class of special cases and consider some proofs to be elegant or beautiful. There is thus more to the method than the mere logic involved in deduction. There is also less to the objects than their intuitive or instinctive origins may suggest.

It is in fact a distinctive feature of mathematics that it can operate effectively and efficiently without defining its objects.

Points, straight lines, and planes are *not* defined. In fact, a mathematician of today rejects the attempts of his predecessors to define a point as something that has "neither length nor width" and to provide equally meaningless pseudo-definitions of straight lines or planes.

The point of view as it evolved through centuries is that one need not know what things *are* as long as one knows what *statements* about them one is *allowed* to make. Hilbert's famous *Grundlagen der Geometrie* begins with the sentence: "Let there be three kinds of objects; the objects of the first kind shall be called 'points,' those of the second kind 'lines,' and those of the third 'planes.' " That is all, except that there follows a list of initial statements (axioms) that involve the words "point," "line," and "plane," and from which other statements involving these undefined words can now be deduced by logic alone. This permits geometry to be taught to a blind man and even to a computer!

This characteristic kind of abstraction, which leads to a nearly total disregard of the physical nature of geometric objects, is not confined to the traditional boundaries of mathematics. Ernst Mach's critical discussion (which owes much to James Clerk Maxwell) of the notion of temperature is a case in point. To define temperature one needs the notions of *thermal equilibrium* and *thermal contact,* but to define these in logically acceptable terms is, at least, awkward and perhaps not even possible. An analysis shows that all one really needs is the *transitivity* of *thermal equilibrium; i.e.,* the postulate. (sometimes called the *zero*th law of thermodynamics) that if (A and B) and (A and C) are in thermal equilibrium, then so are (B and C). For completeness one also needs a kind of converse of the *zero*th law, namely that if A, B, and C are in thermal equilibrium, then so are (A and B) and (A and C). Again, as in geometry, one need not know the (logically) precise meaning of terms, but only how to combine them into meaningful (*i.e.,* allowable) statements.

But while we may operate reliably with undefined (and perhaps even undefinable) objects and concepts, these objects and concepts are rooted in apparent physical (or at least sensory) reality. Physical appearances suggest and even dictate the initial axioms; the same apparent reality guides us in formulating questions and problems.

To exist (in mathematics), said Henri Poincaré, is to be free from contradiction. But mere existence does not guarantee survival. To survive in mathematics requires a kind of vitality that cannot be described in purely logical terms.

In the following chapters we discuss a number of problems that not only have survived but have given birth to some of the most fruitful developments in mathematics. They range from the concrete to the abstract and from the very simple to the relatively complex. They were chosen to illustrate both the objects and the methods of mathematics, and should convince the reader that there is more to pure mathematics than is contained in Bertrand Russell's definition that "Pure mathematics is the class of all propositions of the form 'p implies q,' where p and q are propositions containing one or more variables, the same in the two propositions, and neither p nor q contains any constants except logical constants."

Contents

ix

Mathematics and Logic

Chapter 1 Examples

1. *The Infinity of Primes*

AMONG THE SO-CALLED NATURAL NUMBERS (1, 2, 3, and so on) are some that are divisible only by 1 and by themselves; these are called the *prime numbers.* The prime numbers are the building blocks of all the numbers in the sense that every natural number is the product of powers of the primes that divide it. For instance $60 = 2^2 \cdot 3 \cdot 5$. The first several primes are 1, 2, 3, 5, 7, 11, 13, 17. It can be asked whether the series goes on forever, or, in other words, whether there is a largest prime. The answer is that there is no largest prime. This has been known since the golden age of Greece. It was proved by Euclid in the 3rd century B.C. His argument is as clear and fresh today as ever. Once the infinity of primes is established, a host of other questions about primes arise. Many of these pose real problems and remain unanswered. In this section we shall discuss several of these questions and give Euclid's argument.

We do not know when the notion of a prime integer first appeared and how much time elapsed between the first considerations of the properties of such numbers and the discovery that there are infinitely many of them. Probably after the first tentative considerations and the pragmatic study of such numbers as 2, 3, 11, 17, this question soon arose. The idea of infinity, probably preceded by the notion of "arbitrarily large," must have been entertained for a much longer time; perhaps it arose through contemplation of the physical universe.

The following proof, probably still the simplest, asserts the mere existence of arbitrarily large primes. Suppose the number were finite; there would then be a largest prime p. Consider now the number $n = p! + 1$ ($p!$ is read "p factorial" and equals $1 \cdot 2 \cdot 3 \ldots p$). This number is not divisible by any prime up to p. If there is no prime between p and n, as we have assumed, then n itself would be a prime, contrary to our assumption that p is the largest one.

This remarkably simple and elegant result of Euclid's is one of the first known proofs by contradiction. As is typical of all good mathematics, it settles a question, suggests new ones, and leads to new observations. For example, again using the idea of factorials, we can convince ourselves immediately that there can be arbitrarily long sequences of successive integers, all of which are not primes; *i.e.*, they consist of composite numbers. Given an n, one can find n successive composite numbers by writing: $n! + 2, n! + 3, \ldots, n! + n$. The first of these is certainly divisible by 2, the next by 3, $\ldots$, the last one by n.

If we pursue our example a little further, it may be seen how characteristic it is of mathematical thinking that new problems inexorably arise. They almost always quickly lead to new ones which are difficult and may perhaps even be undecidable.

Granting now that the sequence of primes is infinite, one wants to know more. Can one find their frequency or make an estimate of the number $\pi(n)$ of primes between 1 and a (large) integer n? One can prove that actually $\pi(n)$ is asymptotically $n/\log n$, which means that the ratio of $\pi(n)$ to $n/\log n$ gets closer and closer to 1 as n gets larger and larger. This is the famous prime number theorem, first proved in 1896 by J. Hadamard and by C. J. de la Vallée Poussin. The first proofs involved rather sophisticated notions of mathematical analysis, namely the theory of analytic functions. Only in more recent years was a more elementary (though long and complicated) proof found by P. Erdös and A. Selberg. This proof uses only combinatorial and arithmetical notions and does not require any knowledge of analytic functions.

All the above refers to the *number* of primes relative to the sequence of all integers. The most elementary curiosity will immediately be aroused by other observations. Each even integer studied has been found to be representable as a sum of two primes. The mathematician C. Goldbach (1690–1764) conjectured the general truth of this observation, asserting that *every* even integer is representable as the sum of two primes. To this day his conjecture remains unproved. It has been found to hold for even integers up to 100,000,000. Using electronic computers one could even assemble statistics showing in how many different ways it can be done for each even number $2n$; the number of ways grows rather rapidly with n. In recent times the Soviet mathematician I. M. Vinogradow proved that every sufficiently great odd integer can be represented as a sum of 3 primes!

There is no known formula or expression that allows us to *write down* arbitrarily large primes. We know arithmetical expressions that contain a great number of primes. For example, Euler's formula $N = x^2 + x + 41$ yields different prime numbers for $x = 0, 1, 2, \ldots, 39$, but we do not know whether there are infinitely many integers x for which $x^2 + x + 41$ is a prime. One does not even know whether there exists a polynomial in x that gives infinitely many primes for integral values of x. While there exist such polynomials of first degree (*e.g.*, $2x + 1$), to this day nobody knows whether there exist such polynomials of degree greater than 1.

Neither do we know if there exist infinitely many twin primes, that is, primes that differ by two (*e.g.*, 11, 13; 29, 31; and so on).

These examples typify every part of mathematics, indicating that questions which arise almost automatically are often extremely hard to answer; even though in formal structure they seem not to go very far beyond the established body of knowledge.

There is a great deal known beyond the statements made above about the sequence of primes, however. For example, it is known that there are infinitely many primes of the form $4k + 1$ and of the form $4k + 3$. More generally, it is known that in every arithmetic progression $a \cdot k + b$, where a, b are relatively prime integers and $k = 1, 2, 3, \ldots$, there are infinitely many primes and that asymptotically they have the "right" frequency (*i.e.*, $\dfrac{1}{\varnothing(a)}$) in the set of all the primes.[1] Our examples and comments here are to illustrate one thread of mathematical thought continuing and ramifying through history: mathematicians who observe the properties of a finite collection of numbers (the first and the most fundamental of mathematical objects) find that they desire to establish the observed properties for an infinity or totality of these entities. From the beginning, a characteristic of mathematics was the mathematician's urge to generalize. L. Kronecker's dictum that integers were created by God and everything else in mathematics is man-made expressed this point of view in an extreme form. One may dispute it since simple geometric objects surely have a claim on the ultimate in simplicity and primitivity.

2. *Irrationality of* $\sqrt{2}$

In the system $1, 2, 3, \ldots$ of the so-called *natural numbers* it is not always possible to subtract: $3 - 5 = -2$ is not a natural number because it is negative. In the enlarged system of *integers* $0, \pm 1, \pm 2, \ldots$, it is not always possible to divide: $2 \div 3 = 2/3$ is not an integer. In the further enlarged system of all fractions (*rational numbers*), division (except by zero) is always possible. In this section it will be shown that the rational number system is still not "rich" enough for all arithmetic purposes; namely, that it is not always possible to take roots within the system. For example, this means that the square roots of some rational numbers are not rational. We shall show that $\sqrt{2}$ is not rational. Such "irrationalities" have been studied since ancient Greek times and have stimulated important developments, some of which we shall discuss in this section. Irratio-

---

[1] This means that if $\pi_a(n)$ denotes the number of primes not exceeding n in the progression $ak + b$, then

$$\lim_{n \to \infty} \frac{\pi_a(n)}{\pi(n)} = \frac{1}{\phi(a)}$$

where $\phi(a)$ denotes the number of integers less than a and relatively prime to a.

nality may be understood in terms of the decimal expansion of numbers as follows. If the decimal expansion of a number terminates (as in $1/4 = .25$) or repeats (as in $1/3 = .3333 \ldots$), then it follows that the number is a fraction or rational number; in the contrary case the number is called *irrational*. Of course one would have to know the entire infinite decimal expansion of a number to tell whether it repeats or terminates, and clearly one cannot. Therefore other methods, such as the one we shall use, have to be employed.

There are two kinds of irrational number. Some are roots of algebraic equations (for instance, $\sqrt{2}$ is the root of $x^2 - 2 = 0$) and accordingly are called *algebraic* numbers. A number that is not the root of any algebraic equation with rational coefficients is called *transcendental* because it transcends the operations of ordinary arithmetic. Examples are π and e, the base of the system of natural logarithms. The problem of deciding whether a number is rational, algebraic, or transcendental is in general unsolved.

The class of all rational and irrational numbers together constitutes the *real number* system. To solve all possible algebraic equations (in particular, $x^2 + 1 = 0$), it is necessary to enlarge the real number system by the adjunction of the so-called imaginary number $i = \sqrt{-1}$. The system of all numbers of the form $a + ib$, a and b real, is the *complex number* system. These matters will be treated more fully in Section 7.

Much of mathematics has always centred around proofs of *impossibility* of certain constructions and on finding the limitations of theories and methods. This led repeatedly to enlargements of existing mathematical notions, to extension of systems of axioms, and to introduction of new entities. We will try to illustrate this on perhaps the earliest such example: the proof of the irrationality of $\sqrt{2}$.

Is the length of the diagonal of a square whose side is 1 expressible as a ratio of two integers? In other words, do there exist two integers a,b with no divisor in common such that $(a/b)^2 = 2$? Here is the beautifully simple Greek proof that this cannot be so: For if $a^2 = 2b^2$, then a must be even since $2b^2$ is even, and the square of an odd integer is odd. Therefore, $a = 2a_1$ where a_1 is again an integer. Thus $2a^2 = 4a_1^2 = 2b^2$, and we conclude by dividing by 2 that b must be even or $b = 2b_1$. But, we have assumed that the fraction (a/b) already has been brought to its simplest form; *i.e.*, that a and b had no common factor. This contradiction shows the *impossibility* of representing $\sqrt{2}$ as a fraction and necessitates *defining* it as a different sort of number.

Although the irrationality of $\sqrt{2}$ is well known, the statement can be rephrased in a way that is so surprising as to appear nearly paradoxical.

Consider all rational numbers in the interval from 0 to 1, excluding 0. Each number can be written in a unique way as a fraction a/b where a and b have no divisors in common. Imagine now a/b as a centre of an interval of length

$1/2b^2$; in other words, *cover* a/b by the interval with endpoints $a/b - 1/4b^2$ and $a/b + 1/4b^2$. Since the rational numbers form a dense set (*i.e.*, in every interval no matter how small there are always rational numbers) and, since the sum of lengths of all covering intervals is found to be infinite, it would seem that, having so generously covered all rational numbers, we have automatically covered all numbers. However, we shall show that $\sqrt{2}/2$ remains uncovered! In fact the number $|b^2 - 2a^2|$ being an integer must be at least 1; it cannot be 0 since $\sqrt{2}$ is irrational.[2]

Hence

$$\frac{|b^2 - 2a^2|}{2b^2} \geqslant \frac{1}{2b^2}$$

$$\left|\frac{\sqrt{2}}{2} - \frac{a}{b}\right|\left(\frac{\sqrt{2}}{2} + \frac{a}{b}\right) \geqslant \frac{1}{2b^2}$$

$$\left|\frac{\sqrt{2}}{2} - \frac{a}{b}\right| \geqslant \left(\frac{1}{2b^2}\right)\left(\frac{1}{\frac{\sqrt{2}}{2} + \frac{a}{b}}\right) > \left(\frac{1}{2b^2}\right)\frac{1}{2} = \frac{1}{4b^2}$$

and the assertion that $\sqrt{2}/2$ is not covered follows.

The original Greek proof of the irrationality of $\sqrt{2}$ could be applied, with suitable modifications, to establish that $\sqrt{5}$ (and so on) are irrational. Quadratic irrationalities can be obtained by geometric constructions, but the Greeks already had speculated beyond this. One of their oldest problems was whether $\sqrt[3]{2}$ can be so constructed (the so-called Delic problem of doubling the cube). Centuries passed before the impossibility of this construction by means of a ruler and a compass could be demonstrated.

Again, as in our first example, new questions arose. For example, are there real numbers that are not obtainable as roots of algebraic equations with integer coefficients? Not until the 19th century was the nonalgebraic character (transcendentality) of e and of π first demonstrated.

J. Liouville first described a way to construct real numbers that are transcendental. It remained for the development of set theory by Cantor to establish that "most" real numbers are not algebraic. In fact, the totality of algebraic numbers forms only a denumerable set (*i.e.*, all these numbers can be written down in a single sequence $a_1, a_2, a_3, \ldots$), whereas the totality of all real numbers cannot be so arranged. The reader will find the proof of this assertion in Section 4 of this chapter.

There exist easily definable numbers whose rationality or irrationality has not yet been determined. Euler's constant is one of these, and may be defined as follows: Consider the series $1/2 + 1/3 + 1/4 + \cdots + 1/n + \cdots$. The

[2] As usual, $|x|$ denotes the absolute value of x; *i.e.*, $|-5| = 5$, $|2| = 2$, and so on.

*n*th partial sum of this series (*i.e.*, the sum of the first *n* terms) is about log *n*. The difference between this sum and log *n* tends, with increasing *n*, to a limit usually denoted by *C* (Euler's constant) and approximately equal to 0.6. Despite attempts of many mathematicians, it remains an open question whether *C* is a rational fraction or not; probably it is not even algebraic!

Quadratic irrationalities can be represented as continued fractions of very special form. Any real number can be written as

$$a_0 + \cfrac{1}{a_1 + \cfrac{1}{a_2 + \cfrac{1}{a_3 + \ldots}}}$$

where the *a* values are all integers. For quadratic irrationalities (*e.g.*, numbers like $\sqrt{2}$; numbers of the form $p + q\sqrt{r}$, where *p,q,r* are rational numbers) the sequence of the *a* values is periodic and *vice versa*. In a periodic array of such values, they will all be bounded; that is, given a quadratic irrationality *x*, the *a* values in the development of this number do not exceed a specific fixed bound (that depends on *x*). It is not known whether there exists a single algebraic number of order greater than 2 (*e.g.*, a cubic or higher irrationality) for which the *a* values will be bounded; recently the transcendental character of some numbers (*e.g.*, $\sqrt{2}^{\sqrt{2}}$) has been demonstrated.

3.   *Approximation by Rational Numbers*

It is a familiar practice to approximate irrational numbers by rational numbers. For instance, $\sqrt{2}$ is approximately $1.4 = \frac{140}{100}$, and π is approximately $3.14 = \frac{314}{100}$. Thus every real number can be approximated arbitrarily well by rational numbers. In the decimal expansion of a number, this may be done by adding terms in the expansion. But much more precise and general information about the manner of the approximation can be given using only elementary tools; we shall do so in this section. By using this information, we actually shall produce a transcendental number.

A great deal of penetrating and elegant mathematical work has been devoted to the *approximation* of algebraic and other numbers by rational fractions. How well can one approximate, say $\sqrt{2}$, by a fraction *a/b*? One can do it with arbitrary precision; any real number is a limit of a sequence of rational numbers. However, one wants to do it "economically"; that is, make the $\epsilon = |\sqrt{2} - a/b|$ as small as possible, making values of *b* as small as possible. One can make ϵ smaller than c/b^2 where *c* is a specific constant. However, the quadratic irrationalities are found to be the most difficult of all numbers to approximate by fractions with precision given by a constant divided by b^2.

What was known to the Greeks as the golden number, $\dfrac{\sqrt{5}-1}{2}$, because it is derived from the division of a line segment into extreme and mean ratio (the so-called golden section), is among the most difficult to approximate; the value of c is largest for this number.

Yet some transcendental numbers allow very good approximations in the above sense, a fact used by Liouville in his construction of transcendental numbers bearing his name. Liouville first proved the following theorem:

If α is a root of an irreducible algebraic equation with integer coefficients of degree $n \geqslant 2$, then there is a positive constant γ that depends *only* on α such that for *all* integers p,q one has

$$\left| \alpha - \frac{p}{q} \right| > \frac{\gamma}{q^n} \qquad (\gamma > 0)$$

The proof is simple but requires elements of calculus.

Let

$$f(x) = a_0 x^n + a_1 x^{n-1} + \cdots + a_n$$

be the irreducible polynomial with integer coefficients ($a_0 \neq 0$ since the degree of the equation is n), one of whose roots is α.

The derivative $f'(x)$ is bounded in the interval $[\alpha - 1, \alpha + 1]$; *i.e.*, there is a number M such that

$$|f'(x)| \leqslant M, \qquad \alpha - 1 \leqslant x \leqslant \alpha + 1$$

It is sufficient to consider only rational numbers p/q that lie in the interval from $\alpha - 1$ to $\alpha + 1$. Now

$$\left| f\left(\frac{p}{q} \right) \right| = \frac{|a_0 p^n + a_1 p^{n-1} q + \cdots|}{q^n} \geqslant \frac{1}{q^n}$$

since $f\left(\dfrac{p}{q} \right) \neq 0$ (the polynomial is irreducible) and $|a_0 p^n + a_1 p^{n-1} q + \cdots|$ is an integer.

Using the Mean Value Theorem of differential calculus, we infer that there is an x between α and $\dfrac{p}{q}$ (and hence in the interval from $\alpha - 1$ to $\alpha + 1$) such that

$$f(\alpha) - f\left(\frac{p}{q} \right) = \left(\alpha - \frac{p}{q} \right) f'(x)$$

and hence

$$\frac{1}{q^n} \leqslant \left| f\left(\frac{p}{q} \right) \right| = \left| f(\alpha) - f\left(\frac{p}{q} \right) \right|$$

$$= \left| \alpha - \frac{p}{q} \right| \left| f'(x) \right| \leqslant M \left| \alpha - \frac{p}{q} \right|$$

or

$$\left| \alpha - \frac{p}{q} \right| \geqslant \frac{1}{M} \frac{1}{q^n}$$

which concludes the proof of the theorem.

Consider now the number

$$\alpha = \frac{1}{10} + \frac{1}{10^{2!}} + \frac{1}{10^{3!}} + \frac{1}{10^{4!}} + \cdots$$

$$= \frac{1}{10} + \frac{1}{10^2} + \frac{1}{10^6} + \frac{1}{10^{24}} + \cdots$$

$$= .110001000000000000000000010 \ldots$$

One verifies that

$$0 < \alpha - \left(\frac{1}{10} + \frac{1}{10^{2!}} + \cdots + \frac{1}{10^{m!}} \right) < \frac{2}{10^{(m+1)!}}$$

so that there is a sequence of integers p_m such that

$$0 < \alpha - \frac{p_m}{10^{m!}} \leqslant 2 \left(\frac{1}{10^{m!}} \right)^{m+1}$$

In other words there is a sequence of rational numbers p_m/q_m where $(q_m = 10^{m!})$, such that

$$0 < \alpha - \frac{p_m}{q_m} < \frac{2}{q_m^{m+1}}$$

If α were an algebraic number, then for some *fixed* n we would have

$$\left| \alpha - \frac{p_m}{q_m} \right| > \frac{\gamma}{q_m^2}$$

for *all* m. This would imply that for *all* m

$$\frac{\gamma}{q_m^n} < \frac{2}{q_m^{m+1}}$$

which is impossible if m is sufficiently large. Thus the number α above is transcendental.

In connection with approximating irrational numbers by those that are rational, one should also mention the following important theorem:

If α is irrational, there are *infinitely many* rational numbers p/q (p and q have no divisors in common) such that

$$\left| \alpha - \frac{p}{q} \right| \leqslant \frac{1}{q^2}$$

The proof is interesting because it employs Dirichlet's widely applicable pigeonhole principle; this states that if m objects are distributed in n pigeonholes, and if $m > n$ then at least one pigeonhole must contain at least two objects.

The best-known application of this principle is the proof that at least two people in any sufficiently populous city have exactly the same number of hairs on their heads. In the case of New York City all one needs to know is that the number of hairs on any head is less than the city's population of roughly 8,000,-000. (A person would collapse under the weight of 8,000,000 hairs.) If each person is tagged by his specific number of hairs, at least two people must be tagged by the same number (*i.e.,* have the identical number of hairs).

A somewhat more sophisticated example shows that a city of 21,000 or more must have at least two inhabitants with the same initials. (We assume that a person has either two or three initials.) Since there are 26 letters in the English alphabet, there are 26 × 26 sets of different two-letter initials (A.A., A.B., . . . , Y.Z., Z.Z.) and 26 × 26 × 26 three-letter initials (A.A.A., A.A.B., . . . , Y.Z.Z., Z.Z.Z.).

The total number of different initials thus is

$$(26 \times 26) + (26 \times 26 \times 26) < 21{,}000$$

and consequently, thinking of initials as pigeonholes into which inhabitants of the town are placed, we conclude that at least two of the inhabitants must have the same initials.

Returning to the original problem, let Q be a positive integer, and consider the numbers $0, (\alpha), (2\alpha), \ldots, (Q\alpha)$, where (α) denotes the fractional part of α; *e.g.*, $(5/3) = 2/3$, $(3) = 0$, $(\sqrt{5}) = \sqrt{5} - 2$, etc.

Consider now Q intervals (pigeonholes), $0 \leqslant x < \dfrac{1}{Q}, \dfrac{1}{Q} \leqslant x < \dfrac{2}{Q}, \ldots,$ $\dfrac{Q-1}{Q} \leqslant x < 1$, into which the above $Q + 1$ fractional parts must fall. There must thus be at least one interval containing at least two fractional parts; in other words, there are two distinct positive integers q_1 and q_2, both not exceeding Q, and a positive integer s such that

$$\frac{s}{Q} \leqslant (q_1\alpha) < \frac{s+1}{Q}$$

and

$$\frac{s}{Q} \leqslant (q_2\alpha) < \frac{s+1}{Q}$$

We may as well assume that $q_2 > q_1$ and set $q = q_2 - q_1$ so that $0 < q \leqslant Q$. It follows directly that $q\alpha$ is an integer plus or minus a positive fraction not exceeding $1/Q$; in other words, there exists an integer p such that

$$\left| q\alpha - p \right| < \frac{1}{Q}$$

Thus

$$\left| \alpha - \frac{p}{q} \right| < \frac{1}{Qq} \leqslant \frac{1}{q^2}$$

(since $q \leqslant Q$).

If there were only a finite number of fractions $p_1/q_1, p_2/q_2, \ldots, p_r/q_r$ such that

$$\left| \alpha - \frac{p_i}{q_i} \right| \leqslant \frac{1}{q_i^2} \quad i = 1, 2, \ldots, r$$

then since α is irrational, there is an integer Q such that

$$\left| \alpha - \frac{p_i}{q_i} \right| > \frac{1}{Q}$$

Repeating the above argument we can find an irreducible fraction p/q such that

$$\left| \alpha - \frac{p}{q} \right| < \frac{1}{Qq} \leqslant \frac{1}{Q}$$

Hence p/q cannot be one of the fractions p_i/q_i, where $i = 1,2, \ldots r$. Yet

$$\left| \alpha - \frac{p}{q} \right| < \frac{1}{Qq} \leqslant \frac{1}{q^2}$$

contradicting the assumption that the fractions p_i/q_i, where $i = 1,2, \ldots, r$, were *all* the rational numbers satisfying the above qualifications.

The uses of the pigeonhole principle illustrate very well the nature of mathematical creativity and inventiveness. The principle itself could be derived in due time by a computer starting from axioms of arithmetic. But would the computer recognize the *pertinence* of the principle to, say, the problem of inhabitants with the same initials? If it could, then replacement of humans by computers would become feasible.

4. Transcendental Numbers: Cantor's Argument

The concept of infinity was given a precise mathematical formulation by Georg Cantor, whose work was so surprising as to be unacceptable to mathematicians for a time. Using his ideas it is possible to show, for instance, that transcendental numbers must exist *without actually exhibiting one of them.* We shall give his argument in this section. The main concept is that of a *denumerable* set. Such a set has elements that can be labeled exhaustively by the natural numbers 1,2,3, . . . , and thus enumerated. Though it may appear at first glance that every set is denumerable, this is not so, as we shall see.

In this discussion, as contrasted with that of Section 3, we encounter the difference between *existential* and *constructive* arguments. An important philosophical schism in mathematics developed around this difference, and it will be explored in Chapter 2.

We have already mentioned that Cantor proved the existence of transcendental numbers by showing that the set of all algebraic numbers is *smaller* than the set of all real numbers. Since the argument involves comparing *infinite sets,* and since it proved to be enormously stimulating and fruitful, we shall review it briefly.

To define in precise terms what is meant by two sets being *equally numerous,* one needs the concept of *one-to-one correspondence between sets.* Such a correspondence is simply a *pairing off* of elements of one set with those of the other. In other words, it is a way of associating with each element of one set *one and only one* element of the other. If such a correspondence between two sets can be established, they are said to be equally numerous or of the same power.[8]

An infinite set is said to be *countable* or *denumerable* if it is of the same power as the set of positive integers 1,2,3, ... In other words, a set is countable if its elements can be arranged in a sequence.

It may be proved that a countable union of finite or countable sets is at most countable. (It could be finite if some sets in the union were empty.) It follows that the set of algebraic equations of a *given* degree is countable (remember that the coefficients are integers), and hence the set of algebraic equations of *all* degrees is countable.

Suppose now that the set of all real numbers is countable.

Since each real number can be written *uniquely* as a nonterminating decimal (*e.g.,* 1.1 written as 1.0999 . . .), we can imagine *all* real numbers arranged in a sequence

$$c_1 . c_{11} \ c_{12} \ c_{13} \ \ldots$$
$$c_2 . c_{21} \ c_{22} \ c_{23} \ \ldots$$
$$c_3 . c_{31} \ c_{32} \ c_{33} \ \ldots$$
$$\ldots \ldots \ldots \ldots \ldots$$

where the c values are digits.

Let now d_1 be any digit different from c_1, d_2 any digit different from c_{21}, d_3 different from c_{32}, and so on. The real number

$$d_1 . d_2 \ d_3 \ \ldots$$

is thus different from all numbers of our sequence, and we have a contradiction. Thus the set of all real numbers *cannot* be arranged in a sequence and hence is not countable.

The contrast between the methods of Liouville and Cantor is striking, and these methods provide excellent illustrations of two vastly different approaches toward proving the *existence* of mathematical objects. Liouville's is purely *constructive;* Cantor's is purely *existential.* In the first approach the object is exhibited by an explicit construction; in the second, one proves that the object is an element of a nonempty set.

The introduction of purely existential proofs based on the theory of infinite sets has had a profound effect on the development of mathematics. Perhaps the

---

[8] It should be noted that counting consists merely in establishing a one-to-one correspondence between the objects that are being counted and elements of a *standardized* set. Thus counting on fingers amounts to establishing a correspondence between the set of one's fingers and the elements of the set of objects.

only serious methodological change since the Greeks, it produced possibly the only serious division on philosophical grounds among mathematicians.

To appreciate how far-reaching were the implications of Cantor's method, consider briefly the following remarkable paradox. Since the set of numbers that can be defined in a finite number of words is clearly countable, and since we have shown that the set of all real numbers is not countable, there must be real numbers that *cannot* be defined in a finite number of words.

This striking conclusion perturbed and even shocked many mathematicians. The great Poincaré found it reason enough to fight Cantorism, casting his vote against it in a famous phrase: "... n'envisager jamais que les objets susceptibles d'être défini dans un nombre fini des mots" ("... never consider objects except those that can be defined in a finite number of words").

5. *More Proofs of Impossibility*

The exactness of mathematics is well illustrated by proofs of impossibility. In this section we shall give a number of examples beginning with the classical problem of *doubling the cube;* that is, given a cube, construct another of twice the given cubic volume. If the given cube has side a, then the second cube must have side b such that $b^3 = 2a^3$ or $b = \sqrt[3]{2}a$. Thus the problem can be solved if $\sqrt[3]{2}$ can be constructed. This, as we shall rigorously show, is impossible; the problem goes back to the time of Plato.

We conclude with several elementary but incisive examples of mathematical impossibility in some problems on geometric arrangement.

The unique and peculiar character of mathematical reasoning is best exhibited in proofs of impossibility. When it is asserted that doubling the cube (*i.e.,* constructing $\sqrt[3]{2}$ with a ruler and a compass) is impossible, the statement does not merely refer to a *temporary* limitation of human ability to perform this feat. It goes far beyond this, for it proclaims that *never,* no matter what, will anybody ever be able to construct $\sqrt[3]{2}$ or to trisect a general angle if the only instruments at his disposal are a straightedge and a compass.

No other science, or for that matter no other discipline of human endeavour, can even contemplate anything of such finality. No wonder that even today, through inability or unwillingness to understand what is involved, some people keep trying to double the cube, to trisect an angle, or to square the circle.

Let us explain in some (though not complete) detail what is involved. We start in a plane with a segment of a straight line with length arbitrarily taken as unity. It is a routine matter with a straightedge and a compass to construct segments of lengths that are rational numbers p/q (p,q being integers with only 1 as a common divisor). We shall say that a number α is constructible if one can construct a segment of length α. If we allow directed segments, we can easily extend the concept of constructibility to negative numbers.

If α and β are constructible, then it may be seen that $\alpha + \beta$, $\alpha - \beta$, $\alpha\beta$, and α/β are constructible. In addition, $\sqrt{\alpha\beta}$ is also constructible as shown on fig. 1.

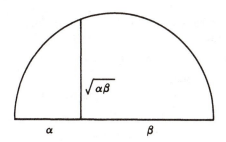

Fig. 1.   The figure is a semicircle of diameter $\alpha + \beta$.

Though we shall not prove it here, the set of constructible numbers is defined by the following two statements: it contains the rational numbers; and if it contains α and β, then it also contains $\alpha \pm \beta$, $\alpha\beta$, α/β, and $\sqrt{\alpha\beta}$. The assertion that duplicating the cube is impossible simply states that $\sqrt[3]{2}$ does not belong to the set of constructible numbers.

Since it will contain the germ of a general proof, let us prove the much weaker statement that $\sqrt[3]{2}$ cannot be written as

$$\frac{a + b\sqrt{c}}{d + e\sqrt{f}}$$

where a,b,c,d,e,f are rational numbers.

We may assume as well that $\sqrt{f}$ and $\sqrt{c}$ are *rationally independent* (*i.e.*, $\sqrt{f}$ is not expressible in the form

$$\frac{\alpha + \beta\sqrt{c}}{\gamma + \delta\sqrt{c}}$$

with rationals α, β, γ, δ); otherwise we could simplify our expression so that it would contain only $\sqrt{c}$.

Suppose now that

$$\sqrt[3]{2} = \frac{a + b\sqrt{c}}{d + e\sqrt{f}}$$

If we rationalize the denominator (*i.e.*, multiply the numerator and the denominator by $d - e\sqrt{f}$), we obtain

$$\sqrt[3]{2} = A + B\sqrt{c} + C\sqrt{f} + D\sqrt{cf}$$

where A, B, C, D are again rational numbers (*e.g.*, $A = \dfrac{ad}{d^2 - e^2 f}$). (Note that $\sqrt{cf}$ is irrational; otherwise $\sqrt{c}$ and $\sqrt{f}$ would fail to be rationally independent.) Write now

$$\sqrt[3]{2} = M + N\sqrt{f} \qquad (N = C + D\sqrt{c}) \qquad (M = A + B\sqrt{c})$$

and cube both sides to obtain

$$2 = (M^3 + 3N^2 Mf) + (3M^2 N + N^3 f)\sqrt{f}$$

We claim now that the coefficient of $\sqrt{f}$ in the expression above vanishes; *i.e.*,

$$3M^2 N + N^3 f = 0$$

Indeed, if it were not so, we could solve for $\sqrt{f}$ to get

$$\sqrt{f} = \frac{2 - M^3 - 3N^2 Mf}{3M^2 N + N^3 f}$$

and $\sqrt{f}$ would be rationally expressible in terms of $\sqrt{c}$ which contradicts the assumption that $\sqrt{c}$ and $\sqrt{f}$ are rationally independent.

Since

$$3M^2 N + N^3 f = 0$$

it follows that

$$(M - N\sqrt{f})^3 = 2$$

and hence both $M + N\sqrt{f}$ and $M - N\sqrt{f}$ are roots of the cubic equation

$$x^3 - 2 = 0$$

Since the sum of the (three) roots of the equation is 0, it follows that $-2M$ is also a root; *i.e.*,

$$(A + B\sqrt{c})^3 = -\tfrac{1}{4}$$

We conclude just as above that both $A + B\sqrt{c}$ and $A - B\sqrt{c}$ are roots of the equation

$$x^3 + \tfrac{1}{4} = 0$$

and hence so is $-2A$.

Thus

$$(A)^3 = \frac{1}{32}$$

so that

$$A = \frac{\sqrt[3]{2}}{4}$$

implying that $\sqrt[3]{2}$ is a rational number. That this is not so can be demonstrated in exactly the way used to show that $\sqrt{2}$ is not rational.

Having thus assumed that $\sqrt[3]{2}$ was of the form

$$\frac{a + b\sqrt{c}}{d + e\sqrt{f}}$$

with a,b,c,d,e,f rational, we have reached a contradiction; hence $\sqrt[3]{2}$ cannot have the assumed form.

Let us now sketch how the above proof can be extended to prove that $\sqrt[3]{2}$ is not constructible. We begin by observing that every constructible number is of the form

$$\frac{a_0 + a_1\sqrt{P_1} + \cdots + a_k\sqrt{P_k}}{b_0 + b_1\sqrt{Q_1} + \cdots + b_l\sqrt{Q_l}}$$

The a and b values are rational numbers, and each P (and Q) is in turn a linear combination with rational coefficients of radicals.

We now introduce the notions of *degree* and of *order* of a constructible number. First, we define the degree of a radical of the form $\sqrt{P}$ (or $\sqrt{Q}$). We say that $\sqrt{P}$ is of degree n if P is a linear combination (with rational coefficients) of radicals of degree $n - 1$ and lower, and if at least one of the radicals is of degree $n - 1$.

For example,

$$\sqrt{\frac{1}{2} + 2\sqrt{\frac{1}{3} + \sqrt{2 + \frac{1}{5}\sqrt{2 + \sqrt{3}}}}}$$

is of degree 5 (rational numbers by themselves are of degree 0).

The degree of a constructible number is simply the largest of the degrees of $\sqrt{P_1}, \ldots, \sqrt{P_k}, \sqrt{Q_1}, \ldots, \sqrt{Q_l}$. It is understood that, in computing the degree of $\sqrt{P}$ (or $\sqrt{Q}$), all radicals are simplified as far as possible. The degree of

$$\sqrt{1 + \sqrt{2 + \sqrt{9}}}$$

for example, is not 3 as it may appear at first glance but 2 since $\sqrt{9} = 3$, and

$$\sqrt{1 + \sqrt{2 + \sqrt{9}}} = \sqrt{1 + \sqrt{5}}$$

If the degree of a constructible number is n, then its order r is the number of rationally independent radicals of degree n. (Rationally independent radicals are those that cannot be obtained from others by rational operations [addition, subtraction, multiplication, and division].) Thus, for example, the expression

$$\frac{a + b\sqrt{c}}{d + e\sqrt{f}}$$

considered above is of degree $n = 1$ and of order $r = 2$.

If we assume now that $\sqrt[3]{2}$ is constructible of degree n and order r, then, proceeding very much as above, we prove that actually it must be of order $r - 1$. Repeating the process we eventually conclude that the order is 0, so that the degree is lower than n. In this way we finally reach a contradiction by proving that $\sqrt[3]{2}$ is rational.

Technically less complicated, but equally sophisticated and perhaps even more typical, is Hilbert's proof of the theorem of Steiner that one cannot find the centre of a given circle by straightedge alone.

What is a construction by straightedge alone? It is a *finite* succession of steps, each requiring that either a straight line be drawn, or a point of intersection of two lines or of a line and the given circle be found. A straight line can be drawn through two points chosen more or less arbitrarily. For example, a step may call for choosing two arbitrary points on the circumference of the circle and joining them by a straight line. Or it may be drawn through two points, one or both of which have been determined in a previous step in the construction as intersections of lines or a line and the circle. The succession of steps eventually must yield a point that can be proved to be the centre of our circle.

Let the construction be performed in a plane P_1 and imagine a transformation or mapping T of the plane P_1 into another plane P_2 such that:

(a) straight lines in P_1 transform into straight lines in P_2. In other words, if points $p, q, r, \ldots$ lie on a straight line l in P_1, their "images" $T(p), T(q), T(r)$, $\ldots$ lie on a straight line $T(l)$ in P_2.

(b) The circumference C of our circle is transformed into a circumference $T(C)$ of some circle in P_2.

As the steps called for in the construction are being performed in P_1, they are being faithfully copied in P_2. Thus when the construction in P_1 terminates in the centre O of C, the "image" construction *must* terminate in the centre $T(O)$ of the circle $T(C)$.

Therefore if one can exhibit a transformation T satisfying (a) and (b), but such that $T(O)$ *is not* the centre of $T(C)$, then the impossibility of constructing the centre of a circle by ruler alone will be demonstrated.

Such a transformation T is shown in fig. 2 and is called a projection through S or simply a projection. Projections distort distances and may transform ellipses into hyperbolas. But a great body of interesting and important properties of geometric configurations remain unchanged (are invariant) by projections. The study of such properties belongs to projective geometry. While ordinary Euclidean geometry may be thought of as the geometry of the straightedge and

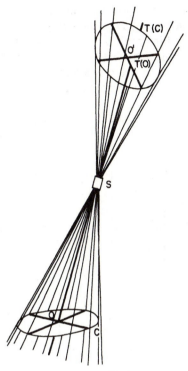

Fig. 2

compass, projective geometry is that of the straightedge alone. Constructions by straightedge alone therefore must be *projective invariants*. On the other hand, the relationship between the circle and its centre is not projectively invariant, thus is not describable in purely projective terms.

For other examples of proofs of impossibility we illustrate combinatorial analysis as a domain of mathematics.

Let a square be subdivided into 64 equal squares to form a chessboard. We omit its lower right-hand and upper left-hand corners. Is it possible to cover the remaining 62 squares by 31 "dominoes" (*i.e.*, rectangles consisting of two adjacent squares)? An elegant proof that this is not possible consists in remarking that if we coloured the squares black and white as on a chessboard, both omitted squares would be of the same colour, say, white. Since each "domino" must cover one black and one white square, no matter how it is laid, it is impossible to cover the remainder, which has two more blacks than whites. The impossibility becomes completely obvious once one thinks to colour the squares, though this is not inherent in the original pattern of a subdivided square.

Another impossibility proof in problems on arrangement of patterns (a part of combinatorial analysis) is the following:

It is possible to decompose a square into a finite number of squares, all of different size.[4] Figure 3 reproduces a simple proof.

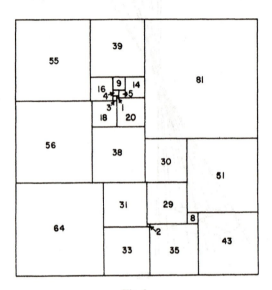

Fig. 3

The question arises whether or not it is possible to have an analogous decomposition for a cube? That is, is it possible to decompose a cube into a finite number of unequal cubes? The answer is in the negative. Assuming that such a decomposition exists, the bottom square would be divided into unequal squares by the bottom faces of the cubes. Consider the smallest of these squares; it could not be located on a corner or on the side of the large square since in this case the two squares adjacent to it would protrude beyond it. There would be no possible square that would fit the fourth side and whose side would be larger than that of our allegedly smallest one. We conclude that the smallest square must be located in the middle of the bottom face of the original cube. Consider now the cubes whose sides are all larger on the four sides of the small square.

---

[4] There is an interesting theory of "squaring the square"; *i.e.*, of finding all possible decompositions of a square into *unequal* squares with parallel sides. This theory is closely allied with Kirchhoff's theory of the flow of electricity in networks. This is another illustration of the remarkable and wholly unexpected connections of which mathematics is full.

They fence off the top face of the cube whose bottom face is our small square. This top face, therefore, has to be covered by a number of cubes of smaller size to cover the top face by squares. Consider the smallest of these and repeat the argument. We will come to the conclusion that the process cannot end in a finite number of steps because, by iterating the argument, we will get smaller and smaller squares on successively higher levels. This denies the possibility of a finite decomposition.

5a. SPERNER'S LEMMA

Another incisive example of mathematical necessity is afforded by Sperner's Lemma. This proposition belongs to that important part of mathematics known as *combinatorial topology,* which classifies geometric objects according to properties that are independent of stretching or smooth distortions generally. For instance, a circle and a square are put into the same class. Technically, this means that one can find a point-to-point correspondence or *mapping* between them that is continuous (meaning that nearby points correspond to nearby points). A famous and remarkable fact about such continuous mappings is the so-called fixed point theorem of Brouwer. This is an immediate consequence of Sperner's Lemma.

Sperner's Lemma, one of the most powerful tools in combinatorial topology, concerns the decompositions of a triangle into smaller triangles and a system of numbering of the vertices.

Imagine a triangle or one of its analogues in higher dimensions; *e.g.,* a tetrahedron in three dimensions. Suppose this triangle (or simplex) is divided "simplicially" into a finite number of smaller triangles. This means the division is such that two of the smaller triangles have a whole side or a vertex in common. Suppose further that we number the vertices of the original triangle 0,1,2 (*see* fig. 4). The vertices of the subdivision that lie on the sides of the original triangle are numbered so that if a point lies on the side 0,1 it has to be marked by 0 or 1; if it lies on the side 1,2, it has to be marked by 1 or 2; and if it is on 0, 2, it has to be marked by 0 or 2. Otherwise the numbering is arbitrary. Here then is Sperner's Lemma: *there must then exist one of the smaller* triangles (sub-simplices) *whose vertices are marked with all the numbers from 0,1,2.* In fact, the number of such "distinguished" sub-simplices is odd!

A similar theorem holds for simplicial decompositions of tetrahedra, this time with numbers 0,1,2,3 to mark the vertices.

Let us first prove the lemma for a simplicial decomposition of an interval; *i.e.,* a decomposition of an interval into subintervals. The endpoints of the original interval are marked 0 and 1, and the interval points of subdivison are marked either 0 or 1. For each subinterval α denote by $\nu(\alpha)$ the number of its endpoints marked 0 so that $\nu(\alpha)$ is 0 (if both endpoints of α are marked 1), 1 (if one end-

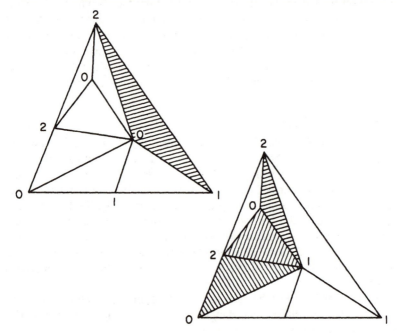

Fig. 4. Two identical simplicial subdivisions with different markings consistent with the requirements of Sperner's Lemma. The shaded triangles are those marked with all the numbers 0,1,2.

point is marked 0 and the other 1), or 2 (if both endpoints are marked 0). A subinterval α is called *distinguished* if one of its endpoints is marked 0 and the other 1; *i.e.*, $\nu(\alpha) = 1$. Let m be the number of distinguished subintervals. Then

$$m - \Sigma\nu(\alpha) = \text{even number}$$

where Σ denotes summation over all the sub-simplices (subintervals).

Note that every internal endpoint marked 0 is counted *twice* in $\Sigma\nu(\alpha)$ (since it is an endpoint of exactly two intervals), and only *one* of the endpoints of the original interval is marked 0. Thus $\Sigma\nu(\alpha)$ is equal to 1 plus an even number and is hence odd. Since m differs from $\Sigma\nu(\alpha)$ by an even number, it follows that m is odd, and the proof of the lemma in the simplest one-dimensional case is complete.

For a triangle we proceed analogously, and for each sub-simplex (triangle) α denote by $\nu(\alpha)$ the number of its sides with endpoints (vertices) marked 0 and 1. A sub-simplex will now be called distinguished if its vertices are marked by all the allowable numbers; *i.e.*, 0, 1, 2. Again let us denote by m the number of distinguished sub-simplices, and also note that $\nu(\alpha)$ is even except when α is distinguished, in which case $\nu(\alpha) = 1$. Thus again

$$m - \Sigma\nu(\alpha) = \text{even number}$$

and again each internal $(0,1)$ side is counted twice; consequently

$$\Sigma\nu(\alpha) = \text{even number} + \text{number of } (0,1) \text{ sides of sub-simplices}$$
$$\text{that lie on the } (0,1) \text{ side of the original triangle}$$

But the number of $(0,1)$ sides of sub-simplices that lie on the $(0,1)$ side of the original triangle is odd because we are dealing with a simplicial decomposition of an interval for which Sperner's Lemma already has been proved. Thus m is odd. Note that Sperner's Lemma for a simplicial decomposition of a triangle reduces to the lemma for the interval; *i.e.*, the *two*-dimensional case is reduced to that of *one* dimension. Similarly, the three-dimensional case (simplicial decomposition of a tetrahedron) can be reduced to one of two dimensions, and the process can be continued indefinitely if we know how to define higher-dimensional simplices.

This very special, purely combinatorial result has far-reaching consequences. We illustrate its applicability by sketching a proof of the famous theorem of Brouwer.

In one dimension the theorem asserts that if an interval is mapped continuously into itself by a mapping T—*i.e.*, to every point p of the interval there corresponds a point $T(p)$ of the same interval, and $T(q)$ can be made arbitrarily close to $T(p)$ by taking q sufficiently close to p—then there exists at least one point p that remains fixed; *i.e.*,

$$T(p_0) = p_0$$

The proof proceeds as follows.

Subdivide the interval simplicially; that is, simply take any number of points in it. Consider two sets of these points. The first set consists of those points whose distance from the left end (the point marked 0) has not decreased after mapping. It certainly contains the point 0 and will all be marked 0. The second set consists of those points whose distance from the right-hand end has not decreased; mark them by 1. Sperner's Lemma asserts that there must be a sub-simplex; *i.e.*, a small interval whose ends are marked by 0 and by 1. This means that there must be two arbitrarily close points such that the distance of one from the left-hand side has not decreased, and the one next to it has not decreased its distance from the right-hand side. They can be made arbitrarily close because our division could have been made arbitrarily fine. Passing to the limit, we conclude that there must be at least one point whose distance from both ends has not decreased; therefore, a point that has not moved at all. In two or more dimensions the argument is essentially the same.

Fixed-point theorems are among the most powerful tools of modern mathematics. The proof given above followed from a "finitistic" consideration concerning the *impossibility* of numbering vertices of sub-simplices in a manner that would not allow at least one complete set of indices on a subsimplex.

Of the last three examples, the problem of dominoes on the chessboard and the problem of dividing a cube into smaller cubes that are all different may seem mere puzzles or curios (perhaps rightly so). Sperner's Lemma turns out to be much more important and profound, with many applications. Formally though, they have something in common; they all concern enumerations of certain patterns.

Very often the number of different applications in seemingly unrelated parts of mathematics contributes to the importance and beauty of a result. In the words of Descartes, "from consideration of examples, one can form a *method*." In general one cannot formally draw a line between the sublime and the ridiculous. The criteria are in part aesthetic and are also governed by applicability and pertinence to other situations in mathematics.

6.   *The Art and Science of Counting*

In this section we consider an elementary problem in counting and find ourselves inescapably involved with complex numbers. These were introduced in Section 2. This illustrates the typically broad ramifications of counting problems.

Counting is such a primitive process, and we are first exposed to it at such an early age, that it may come as a surprise to learn that it is also a source of many problems of great interest, importance, and difficulty.

Let us consider briefly the problem of finding in how many different ways one can change a dollar. In other words, how many different solutions does the equation

$$100 = l_1 + 5l_2 + 10l_3 + 25l_4 + 50l_5$$

have, where by a "solution" we understand a quintuplet of nonnegative integers $(l_1, l_2, l_3, l_4, l_5)$? (Here l_1 represents the number of pennies, l_2 the number of nickels, and so on.)

If one simply tries to enumerate the various possibilities, one soon becomes discouraged by the near hopelessness of the task.

Let us rephrase the problem. Consider the series

$$(1 + x + x^2 + x^3 + \cdots)$$
$$(1 + x^5 + x^{10} + x^{15} + \cdots)$$
$$(1 + x^{10} + x^{20} + x^{30} + \cdots)$$
$$(1 + x^{25} + x^{50} + x^{75} + \cdots)$$
$$(1 + x^{50} + x^{100} + x^{150} + \cdots)$$

where the exponents of x in the first series are multiples of 1, in the second multiples of 5, and so on.

If we multiply these series formally (*i.e.*, disregard that they are infinite in length and treat them as ordinary polynomials), we will obtain a series of the form

$$1 + A_1 x + A_2 x^2 + A_3 x^3 + \cdots$$

and it may be seen that A_m is simply the number of different ways of writing m in the form $l_1 + 5l_2 + 10l_3 + 25l_4 + 50l_5$. In particular A_{100} is the desired number of different ways of constituting a dollar with coins.

Observe that each series is geometric; consequently the above product is equal (again formally) to

$$\frac{1}{(1 - x)\ (1 - x^5)\ (1 - x^{10})\ (1 - x^{25})\ (1 - x^{50})}$$

Let us interrupt the argument for a moment to consider a much simpler problem. Suppose we wanted the number of different solutions of

$$100 = l_1 + 2l_2$$

In this case we would be led to

$$\frac{1}{(1 - x)\ (1 - x^2)} = \frac{1}{(1 - x)^2\ (1 + x)}$$

We could now attempt a decomposition into partial fractions; this means we would try to find numbers a, b, c such that identically in x (*i.e.*, for all x) one has

$$\frac{1}{(1 - x)^2\ (1 + x)} \equiv \frac{a}{(1 - x)^2} + \frac{b}{1 - x} + \frac{c}{1 + x}$$

This leads to the identity

$$1 \equiv a(1 + x) + b(1 - x)\ (1 + x) + c(1 - x)^2$$

which yields three simultaneous linear equations

$$-b + c = 0$$
$$a - 2c = 0$$
$$a + b + c = 1$$

Thus $a = \frac{1}{2}$, $b = \frac{1}{4}$, $c = \frac{1}{4}$; hence

$$\frac{1}{(1 - x)^2\ (1 + x)} = \frac{1}{2}\frac{1}{(1 - x)^2} + \frac{1}{4}\frac{1}{1 - x} + \frac{1}{4}\frac{1}{1 + x}$$

We have

$$\frac{1}{1 - x} = 1 + x + x^2 + x^3 + \cdots$$

$$\frac{1}{1 + x} = 1 - x + x^2 - x^3 + \cdots$$

$$\frac{1}{(1 - x)^2} = 1 + 2x + 3x^2 + 4x^3 + \cdots$$

and the coefficient of x^{100} in the product

$$(1 + x + x^2 + \cdots)(1 + x^2 + x^4 + x^6 + \cdots)$$

is thus

$$\tfrac{1}{2}(101) + \tfrac{1}{4}(1) + \tfrac{1}{4}(1) = 51$$

To solve this problem we did not need all the machinery of series and partial fractions. Here we could enumerate immediately by noting that l_1 must be even (since $100 - 2l_2$ is even) and that there are exactly 51 even integers from 0 to 100 inclusive. However, having solved a simple problem in an unnecessarily complicated way, we are now in a much better position to see what should be done to deal with the more complicated question. Observe that the number of pennies must be divisible by 5; consequently the number of ways of changing a dollar A_{100} is equal to

$$B_{100} + B_{95} + B_{90} + \cdots + B_5 + B_0$$

where B_m is the number of solutions of the equation

$$m = k_1 5 + k_2 10 + k_3 25 + k_4 50$$

or, equivalently, the number of solutions (in nonnegative integers) of

$$\frac{m}{5} = k_1 + 2k_2 + 5k_3 + 10k_4$$

(m is divisible by 5).

It follows that A_{100} is the sum of coefficients of

$$1, x, x^2, \ldots, x^{19}, x^{20}$$

in

$$(1 + x + x^2 + \cdots)(1 + x^2 + x^4 + \cdots)(1 + x^5 + x^{10} + \cdots)(1 + x^{10} + x^{20} + \cdots)$$
$$= \frac{1}{(1-x)(1-x^2)} \frac{1}{(1-x^5)(1-x^{10})}$$

As we have seen

$$\frac{1}{(1-x)(1-x^2)} = \frac{1}{2}\frac{1}{(1-x)^2} + \frac{1}{4}\frac{1}{1-x} + \frac{1}{4}\frac{1}{1+x}$$
$$= 1 + x + 2x^2 + 2x^3 + 3x^4 + 3x^5 + \cdots$$

and substituting x^5 for x we have

$$\frac{1}{(1-x^5)(1-x^{10})} = 1 + x^5 + 2x^{10} + 2x^{15} + 3x^{20} + \cdots$$

Thus

$$(1 + x + x^2 + \cdots)(1 + x^2 + x^4 + \cdots)(1 + x^5 + x^{10} + \cdots)(1 + x^{10} + x^{20} + \cdots)$$
$$= (1 + x + 2x^2 + 2x^3 + 3x^4 + 3x^5 + \cdots)(1 + x^5 + 2x^{10} + 2x^{15} + 3x^{20} + \cdots)$$

Performing the multiplication and adding up the coefficients of powers of x from 0 to 20 inclusive, with a bit of labour we obtain 292; thus there are 292 different ways of changing a dollar!

The idea of using power series (*i.e.,* series of the form $a + a_1 x + a_2 x^2 + \ldots$; speaking very loosely, polynomials of infinite length) to do the counting proved extraordinarily fruitful.

In solving the problem of changing a dollar we made use of special features of the U.S. monetary system. For instance, we knew that, except for pennies, the values of all coins are divisible by 5. We also understood that the ratio of the value of the dime to the nickel is 2, the same as that of the fifty-cent piece to the quarter. Judicious utilization of such incidental properties saved labour by allowing the fullest use of integers. But this tended to hide the full power and generality of the method.

To see what is involved a little better, consider the problem of finding the number A_n of solutions (in nonnegative integers) of the equation

$$n = l_1 + 2l_2 + 3l_3$$

(*n* a nonnegative integer).

As above we are led to find the coefficient of x^n in the expansion of

$$\frac{1}{(1 - x)(1 - x^2)(1 - x^3)}$$

Again we attempt a decomposition into partial fractions, first factoring $(1 - x)(1 - x^2)(1 - x^3)$ to give

$$(1 - x)(1 - x^2)(1 - x^3) = (1 - x)^3 (1 + x)(1 + x + x^2)$$

But to produce linear factors from $1 + x + x^2$ we need *complex numbers.*

In fact

$$x^2 + x + 1 = (\alpha x + 1)(\bar{\alpha} x + 1)$$

where

$$\alpha = \frac{1}{2} + \frac{\sqrt{3}}{2} i, \qquad \bar{\alpha} = \frac{1}{2} - \frac{\sqrt{3}}{2} i$$

and $i^2 = -1$.

The decomposition into partial fractions is of the form

$$\frac{1}{(1 - x)(1 - x^2)(1 - x^3)}$$

$$\equiv \frac{a}{(1 - x)^3} + \frac{b}{(1 - x)^2} + \frac{c}{1 - x} + \frac{d}{1 + x} + \frac{e}{1 + \alpha x} + \frac{f}{1 + \bar{\alpha} x}$$

and a,b,c,d,e,f can be found by solving six simultaneous linear equations.

If we now convince ourselves that

$$\frac{1}{(1 - x)^3} = \frac{1}{2}(1 \cdot 2 + 2 \cdot 3x + 3 \cdot 4x^2 + 4 \cdot 5x^3 + \cdots)$$

then this together with the already noted formula

$$\frac{1}{(1-x)^2} = 1 + 2x + 3x^2 + 4x^3 + \cdots$$

as well as with repeated use of the geometric series identity

$$\frac{1}{1-q} = 1 + q + q^2 + \cdots$$

(with $q = x$, $q = -x$, $q = -\alpha x$, $q = -\bar{\alpha}x$) yields

$$A_n = \frac{a}{2}(n+1)(n+2) + b(n+1) + c$$
$$+ d(-1)^n + e(-1)^n\alpha^n + f(-1)^n\bar{\alpha}^n$$

Although the answer is incomplete since we have not determined the coefficients $a, b, \ldots, f$, their determination is routine and has little bearing on the main points of our discussion.

Perhaps the most striking feature of the solution is the emergence of complex numbers in connection with a problem that involves counting (hence only integers). It therefore seems appropriate to digress briefly into the subject of the nature and the evolution of numbers.

7. Digression on the Number System and on Functions

The scheme of the number systems, sketched in the introduction to Section 2, is discussed here in some detail.

Following Kronecker we take positive integers as God-given (though they can be built up from more primitive set-theoretic and logical notions). We then note that the operations of addition and multiplication on positive integers yield again positive integers; therefore as long as one considers only these two operations the set of positive integers is a closed, self-contained universe of discourse.

But already subtraction forces us to go beyond and to extend this comfortable closed universe. Indeed, as simple an equation as

$$3 + x = 2$$

has no solution in the realm of positive integers.

We introduce 0 and negative integers just to make subtraction always possible. In extending the number system we also extend the operations. In so doing care must be taken so that the usual properties (associativity, commutativity, and distributivity) are preserved. It is because of this that 0 times any number must be taken to be 0, and the product of two negatives comes out positive.[5]

---

[5] Let a be an integer, then $0a = (1-1)a = a - a = 0$ if we insist (as indeed we do) that multiplication and subtraction be distributive. Similarly, $2(1-1) = 0$ and hence $(-1)2 = -2$; $(2+(-2))(-1) = 0$, whence $-2 + (-2)(-1) = 0$ and $(-2)(-1) = 2$.

Having extended the number system to include 0 and negative integers, we find that an equation like

$$ax = b \qquad a \neq 0$$

(where a and b are integers) may not be soluble. We thus again extend the system by introducing fractions with numerators and denominators that are integers (positive and negative). The operations of addition and multiplication are now defined in the usual way, again with an eye to preserving the basic properties (associativity, commutativity, and distributivity) of these operations. Subtraction and division become universally possible (except division by 0), and rational numbers finally form a set that is *closed* under *all four* arithmetic operations. Unfortunately (or perhaps fortunately!) rational numbers do not suffice. The answer to the simple problem of finding the ratio of the length of the hypotenuse of a right isosceles triangle to the length of one of its legs is $\sqrt{2}$ and, as we have seen, $\sqrt{2}$ is not a rational number.

To extend the number system to irrational numbers (like $\sqrt{2}$) was a task of a different nature and of far greater subtlety and difficulty, for it was accomplished only by considering infinitesimal operations.

Roughly speaking, one can proceed as follows: Rational numbers in decimal notation may terminate (*e.g.*, 2.13) or may be of the form

$$a_0.a_1a_2\ldots a_mb_1b_2\ldots b_nb_1b_2\ldots b_n\ldots$$

which is a pattern of digits that repeats itself indefinitely after a specific point. For example

$$\frac{1}{7} = 0.142857\ 142857\ 142857\ \ldots$$

Irrational numbers now can be identified with nonterminating decimals that do not have a repeating pattern.

But a nonterminating decimal $a_0.a_1a_2a_3\ldots$ is simply another way of writing the *infinite* series

$$a_0 + \frac{a_1}{10} + \frac{a_2}{10^2} + \frac{a_3}{10^3} + \cdots$$

and infinite series have been considered tricky ever since Zeno's paradox about Achilles and the tortoise.[6]

Since $0 \leqslant a_m \leqslant 9$ for every m, we must have for $k = 2,3\ldots$

$$\frac{a_{n+1}}{10^{n+1}} + \frac{a_{n+2}}{10^{n+2}} + \cdots + \frac{a_{n+k}}{10^{n+k}} \leqslant \frac{9}{10^n}\left(\frac{1}{10} + \frac{1}{10^2} + \cdots\right)$$

$$\leqslant \frac{9}{10^n}\frac{1}{9} = \frac{1}{10^n}$$

[6]"Zeno (of Elea)," *Encyclopædia Britannica* (1968).

and consequently the infinite series above can be thought of as defining a sequence of nested intervals I_n with endpoints

$$a_0 + \frac{a_1}{10} + \cdots + \frac{a_n}{10^n} \text{ and } a_0 + \frac{a_1}{10} + \cdots + \frac{a_n}{10^n} + \frac{1}{10^n}$$

If we now take the view that an irrational number is *defined* if it can be approximated with arbitrary accuracy by terminating decimals, then indeed the infinite series above, or, equivalently, the sequence of nested intervals, defines a number.

Again the arithmetical operations are appropriately defined, and the real numbers (*i.e.*, rational and irrational numbers) are closed under all four operations (of course, division by 0 is not allowed). On the other hand, as contrasted with rationals, real numbers have the so-called Dedekind property; *i.e.*, if they are divided into two *nonempty* classes A and B such that every real number is either in A or in B, and *every* number in A is less than every number in B, then either A has a largest or B a smallest number. The Dedekind property expresses the fact that real numbers form a *continuum*.

Although a rigorous theory of real numbers was not established until the latter part of the 19th century (mainly by Dedekind and Cantor, the sketch above being close in spirit to Cantor's theory), mathematicians had no qualms about using real numbers freely, if perhaps somewhat uncritically.

But the process of extending the number system was not over. Simple quadratic equations fail to yield real solutions; *e.g.*, $x^2 + 2x + 2 = 0$, when the application of the familiar formula for the roots of a quadratic gives

$$x_1 = 1 - \sqrt{-1} \text{ and } x_2 = 1 + \sqrt{-1}$$

To deal with this situation we introduce symbols of the form $a + bi$, where a and b are real. Addition and subtraction are defined in the obvious ways, and to multiply two such "numbers" we follow the usual rules of algebra, except that at the end we replace i^2 by -1.

Thus

$$(2 - i)(3 + 2i) = 6 - 3i + 4i - 2i^2 = 6 + i - 2(-1) = 8 + i$$

Curiously enough, complex numbers (*i.e.*, symbols of the form $a + bi$ with addition and multiplication defined as above) caused mathematicians some discomfort although they were completely absorbed into the body of mathematics by the early 19th century. Some doubts and fears must linger even now because until very recently secondary school curricula largely avoided the subject of complex numbers.

Perhaps the original name (they were first called imaginary numbers) helped create a slight aura of mysticism. Perhaps it was difficult to depart from the notion that numbers are meant to represent *measurements,* and complex numbers did not readily fit into this way of thinking.

Be that as it may, the extension of the number system to include complex numbers brought untold benefits to mathematics.

Above all, and perhaps most miraculously, having introduced complex numbers merely to make all quadratic equations with *real* coefficients solvable, it turned out that all *algebraic equations* (even those with complex coefficients) became solvable.

This remarkable fact, known as the *fundamental theorem of algebra,* can also be stated as follows: An equation

$$a_n z^n + a_{n-1} z^{n-1} + \cdots + a_0 = 0 \quad a_n \neq 0$$

of nth degree with complex coefficients has n complex roots (that need not all be different). In other words, there exist n complex numbers $z_1, z_2, \ldots, z_n$ such that

$$a_n z^n + a_{n-1} z^{n-1} + \cdots + a_0 \equiv a_n(z - z_1) (z - z_2) \cdots (z - z_n)$$

Thus all polynomials with complex coefficients can yield linear factors; we needed this fact before we could use partial fractions to solve problems in counting. If one finds it striking that a formula for the number A_n of solutions in nonnegative integers l_1, l_2, l_3 of the equation

$$n = l_1 + 2l_2 + 3l_3$$

involves complex numbers, it should not be because of some vague feeling that "imaginary quantities" help solve a concrete and real problem. There is no place in mathematics for such simple-minded mysticism. What is striking is that a concept introduced to make quadratic equations solvable should prove so decisively useful in so unexpected a context.

Complex numbers are conveniently represented as points in the plane. In fact, if we choose once and for all an x-axis and a y-axis, the complex number $z = x + iy$ is represented by the point (x,y).

Once complex numbers were introduced, they stimulated study of *functions of a complex variable; i.e.,* mappings of sets of complex numbers into complex numbers.

The simplest functions of a complex variable exhibit striking properties. For example, the function

$$w = \frac{1 + z}{1 - z}$$

maps the interior of the circle of radius 1 around the origin ($z = 0$) into the half-plane to the right of the y-axis! (See fig. 5.)

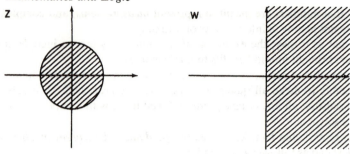

Fig. 5.   The function $w = \dfrac{1 + z}{1 - z}$ maps the shaded region in the z-plane into the shaded half-plane in the w-plane.

Since multiplication by i corresponds geometrically to rotating the plane by 90° counterclockwise, we see that

$$w = i\frac{1 + z}{1 - z}$$

maps the unit circle about the origin into the half-plane above the axis.

As soon as one thinks of functions of a complex variable, one is immediately led to the idea of developing the differential and the integral calculus of such functions. Of special interest are functions that have a derivative; *i.e.*, for which the limit

$$\lim_{\Delta z \to 0}\frac{f(z + \Delta z) - f(z)}{\Delta z}$$

exists for every z in some region. The fact that the limit must exist no matter how the complex "increment" Δz has approached zero imposes such a severe restriction on the function f that as soon as the first derivative exists in a region, derivatives of *all orders* exist in the region. Moreover, around every point z_0 in the region, the function f is representable as a power series; *i.e.*, one has

$$f(z) = a_0 + a_1(z - z_0) + a_2(z - z_0)^2 + \cdots$$

and the series converges in a certain circle with centre z_0. (The radius of such a circle of convergence may be infinite, in which case f is called an *entire function*.) Functions that have a derivative in a region are called analytic in the region, and their theory forms one of the most important and beautiful chapters of mathematics. Analytic functions have found their way into almost every corner of mathematics and physics, from hydrodynamics to number theory and from quantum mechanics to topology.

The evolution of complex analysis from its humble origin in solving quadratic equations to the magnificent edifice of the theory of analytic functions illustrates again the vitality of mathematical concepts.

8. *The Art and Science of Counting* (*Continued*)

We have seen several times that problems of great depth and difficulty lie close to those that are mere puzzles and quite easy to solve.

The problem of finding the number of ways of changing a dollar, or the related problem of finding the number of solutions (in nonnegative integers) of the equation

$$n = l_1 + 2l_2 + 3l_3$$

is quite easy to solve once the idea of using power series (so-called *generating functions*) becomes available.

But of incomparably greater difficulty is the problem of finding the number $p(n)$ of solutions of

$$n = l_1 + 2l_2 + 3l_3 + \cdots$$

i.e., the number of ways of partitioning (or changing) n into smaller numbers. It is the famous problem of *partitio numerorum* of Euler.

In terms of generating functions we have

$$1 + p(1)x + p(2)x^2 = \frac{1}{1-x}\frac{1}{1-x^2}\frac{1}{1-x^3}\cdots$$

and because now we do not restrict the sizes of the portions into which n is to be partitioned, the right-hand side shows an infinite product. We no longer can apply the method of partial fractions in a straightforward way. Only in 1934 did H. Rademacher find a beautiful (but rather complicated) formula for $p(n)$, in one of the most ingenious and subtle uses of complex analysis.

The complex analysis used by Rademacher was an extension of the method by means of which Hardy and Ramanujan (1917) found their justly famous asymptotic formula

$$p(n) \sim \frac{1}{4n\sqrt{3}} \exp(\pi\sqrt{\tfrac{2}{3}n})$$

(It is worth mentioning that $p(n)$ increases very rapidly with n; for example, $p(200)$ is a thirteen-digit number!)

The question of *partitio numerorum* in additive number theory is related to the famous problem of representing integers as sums of squares.

Late in the 18th century Lagrange proved that every positive integer is a sum of four squares (three squares do not suffice; *e.g.*, 19 is not a sum of three squares). This "four square theorem," though largely superseded by the later work of Jacobi, is still counted among the great mathematical achievements.

Jacobi went much further, for he determined the number of different ways a number can be written as a sum of squares. To make the counting easier it is convenient to consider as distinct those representations that differ in order or sign. Thus

$$2^2 + 1^2 + 1^2 + 1^2$$
$$(-2)^2 + (-1)^2 + 1^2 + 1^2$$
$$1^2 + 2^2 + (-1)^2 + 1^2$$

and so on, are all counted as distinct representations of 7 as a sum of four squares.

This convention helps to show that if $r(m)$ is the desired number of representations of m, then

$$1 + r(1)x + r(2)x^2 + r(3)x^3 + \cdots$$
$$= (\ldots x^{(-2)^2} + x^{(-1)^2} + 1 + x^{1^2} + x^{2^2} + \cdots)^4$$
$$= (1 + 2x^{1^2} + 2x^{2^2} + 2x^{3^2} + \cdots)^4$$

By a series of brilliant transformations Jacobi proved the identity

$$(1 + 2x^{1^2} + 2x^{2^2} + 2x^{3^2} + \cdots)^4$$
$$= 1 + 8\frac{x}{1-x} + 8\frac{2x^2}{1-x^2} + 8\frac{3x^3}{1-x^3} + 8\frac{5x^5}{1-x^5} + \cdots$$

where the powers l of x in terms

$$\frac{x^l}{1-x^l}$$

run through all integers that are *not* divisible by 4.

Since

$$\frac{x^l}{1-x^l} = x^l + x^{2l} + x^{3l} + \cdots$$

one finds almost directly that

$r(m) = 8$ multiplied by the sum of divisors of m that are not multiples of 4

Since 1 is a divisor of every integer, and since it is clearly not divisible by 4, one gets for all integers $m > 0$

$$r(m) \geqslant 8$$

which implies Lagrange's theorem (since it merely states that $r(m) > 1$).

The identity of series that gave rise to Jacobi's wonderful formula for $r(m)$ is one of a class of remarkable identities that Jacobi and others have discovered. Some relate *partitio numerorum* to the problem of representing numbers as sums of squares.

The discovery of these identities was not a haphazard or accidental affair; though, as in other disciplines, discoveries in mathematics sometimes are due to luck, and it is the deserving ones who are usually lucky. They were discovered mainly through a systematic development of a theory of a class of analytic functions called *elliptic functions*. Elliptic functions, in turn, first suggested themselves to mathematicians through the problems of finding the length of the circumference of an ellipse (hence the adjective "elliptic") and of describing the motion of a pendulum. Though both problems lead to integrals that belong to real analysis (ordinary calculus), the complex domain was found to be much more appropriate for their study. The benefits of transplanting the problem from its seemingly proper setting to what appeared foreign soil were immense; for only complex analysis made the surprising transition from the motion of the pendulum to representation of integers as sums of squares seem natural and even inevitable. Although it is possible to prove many of the identities in a direct and elementary fashion (*e.g.*, without using complex numbers), the deeper reasons for "what makes them tick" lie in the theory of elliptic functions. The tendency toward economy of thought provided by a theory is just as strong among mathematicians as it is among natural scientists.

9. *Elementary Probability and Independence*

The theory (or calculus) of probabilities has its logical and historical beginnings in simple problems of counting. In an experiment involving a chance outcome (such as drawing a card from a deck), the probability that a given event will occur is taken as the ratio of the number of outcomes yielding the event to the number of possible outcomes. This is how one finds the odds in familiar games of chance. In times past one did little more with probability theory, but in the 20th century the theory has undergone a great development and has become a major part of mathematics, with ramifications and applications not only to other parts of mathematics but to other sciences as well. Yet the logical structure of this theory is remarkably simple. In this section we shall give an account of this theory and its development, and in the following section we shall touch upon some of its more technical ramifications and further developments.

In many experiments (*e.g.*, tossing a coin, rolling a pair of dice, dealing cards) no specific outcome is predictable with certainty. It is usually possible, however, to list *all possible outcomes,* and in many instances there may be a finite (though perhaps staggeringly large) number of them. We may be interested in some subset of outcomes and we may wish to assign, hopefully in a sensible way, a *number* to the event in which a given outcome belongs to this subset.

More abstractly, we may say that to *every subset A* of the set Ω of all outcomes we wish to assign a number $p(A)$ that in some way will be a measure of the likelihood that the outcome will belong to A.

In case Ω is a *finite* set, Laplace proposed to define $p(A)$ as the ratio of the number $\nu(A)$ of elements in A to the total number $\nu(\Omega)$ of elements in Ω; *i.e.*,

$$p(A) = \frac{\nu(A)}{\nu(\Omega)}$$

provided *all outcomes could be considered equiprobable.*

The last proviso makes the definition circular, for the concept of *probability* then is dependent upon the concept of *equiprobability*. From the purely technical point of view, Laplace's definition reduces calculation of probabilities to counting.

Let us illustrate with a typical example how the definition is applied.

Suppose we toss a coin n times; we want the probability that exactly m $(1 \leqslant m \leqslant n)$ heads will show. Let us associate with each toss the letter H if the outcome is heads and the letter T if the outcome is tails. Then the outcome of n tosses can be recorded as a sequence of n letters, each being H or T. If we call such sequences "words," the set Ω of all possible outcomes of n tosses can be thought of as the set of all possible words of length n that have two letters (H and T) only. It may be seen that the total number of such words is 2^n; *i.e.*,

$$\nu(\Omega) = 2^n$$

How many of these words contain H exactly m times? This is a relatively simple problem in counting.

Let us solve it in a way that will illustrate a widely useful method. Let $C(n;m)$ denote the desired number of words and consider $C(n + 1;m)$; *i.e.*, the number of words of length $n + 1$ containing H exactly m times. Those words that end in H can be identified with words of length n containing m H's; and those that end in T with those of length n containing $(m - 1)$ H's. Thus

$$C(n + 1;m) = C(n;m) + C(n;m - 1)$$

and we have what is called a *recursion formula*, one that allows us to reduce the solution of the problem for $(n + 1)$ to the solution of the same problem for n. This method of reduction is called *induction;* another illustration of this method (in a quite different spirit) is the proof of Sperner's Lemma earlier.

The recursion formula above can be used to derive an explicit formula for $C(n;m)$ resulting in

$$C(n;m) = \frac{n!}{(n - m)!m!}$$

The numbers $C(n;m)$ are the familiar binomial coefficients; *i.e.*,

$$(x + y)^n = C(n; 0)x^n + C(n;1)x^{n-1}y + \cdots + \\ C(n; m)x^{n-m}y^m + \cdots + C(n;n)y^n$$

If all outcomes of n tosses can be considered equiprobable, we obtain by Laplace's definition that the probability of getting m heads in n tosses is

$$p(m) = \frac{C(n;m)}{2^n} = \frac{1}{2^n} \frac{n!}{(n-m)!m!}$$

because there are 2^n possible outcomes (heads or tails, compounded n times), and there are $C(n;m)$ ways for m of the tosses to be heads.

Suppose that we toss a die n times and ask for the probability that exactly m 2's will show. The number of possible outcomes is 6^n, and we need only calculate the number of ways in which exactly m 2's can appear.

Let us associate the letter H with 2 and the symbols T_1, T_3, T_4, T_5, T_6 with the remaining numbers. For each of the words of length n containing exactly m H's and which use only letters H and T, there are now 5^{n-m} words of the same length (n) and containing the same number of H's (m) but which use the H and five (5) distinct symbols for T.

Thus the number of favourable outcomes (*i.e.*, outcomes with exactly m 2's) is

$$\frac{n!}{(n-m)!m!} 5^{n-m}$$

Since, as noted above, the total number of outcomes is 6^n, the probability that exactly m 2's will show in n tosses of a dice is

$$\frac{n!}{(n-m)!m!} \frac{5^{n-m}}{6^n} = \frac{n!}{(n-m)!m!} \left(\frac{1}{6}\right)^m \left(\frac{5}{6}\right)^{n-m}$$

Let us now consider a coin that is "loaded" (or lopsided) in such a way that the probability of H in a single toss is $1/6$; the probability of T in a single toss is consequently $5/6$. Suppose we toss this coin n times and again ask for the probability that exactly m H's will show. It is now awkward to describe the equiprobable outcomes unless one resorts to the artifice of thinking of the coin as a six-faced die with one face identified with H and all others with T. The situation becomes even more awkward if the coin is loaded to make the probability of H irrational; *e.g.*, $\sqrt{2}/2$. In such a case one is forced into considering a many-faced die and passing to an appropriate limit as the number of faces becomes infinitely large.

The awkwardness and logical inadequacy of Laplace's definition made mathematicians suspicious of the whole subject of probability. To make matters worse, attempts to extend Laplace's definition to cases in which the number of possible outcomes is infinite resulted in seemingly even greater difficulties. This was dramatized by Bertrand, who considered the problem of finding the probability that a chord of a circle chosen at "random" be longer than the side of an inscribed equilateral triangle.

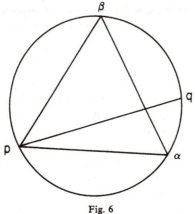

Fig. 6

If we fix one end of the chord, we can think of the circumference of the circle as being the set Ω of all possible outcomes and the arc $\alpha\beta$ on fig. 6 as the set A of "favourable outcomes" (*i.e.*, those resulting in a chord that is longer than the side of an inscribed equilateral triangle).

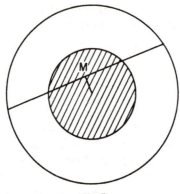

Fig. 7

It thus seems proper to take $1/3$ as the desired probability (*i.e.*, the ratio of the length of the arc $\alpha\beta$ to the total length of the circumference).

On the other hand we can think of the chord as determined by its midpoint M, and thus consider the interior of the circle as being the set Ω of all possible outcomes. The set A of "favourable outcomes" is now the shaded circle on fig. 7 whose radius is one half that of the original. It now seems equally proper to take $1/4$ for our probability, the ratio of the shaded circle's area to that of the original circle.

That two seemingly appropriate ways of solving the problem lead to different answers was so striking that the example became known as "Bertrand's paradox." It is not, of course, a logical paradox. It is simply a warning against uncritical use of the expression "at random." Coming as it did on top of other ambiguities and uncertainties, it greatly helped strengthen the negative attitude toward anything having to do with chance and probability.

Having discussed some of the difficulties, let us describe one of the triumphs of Laplace's theory.

If a coin is loaded so that the probability of H in a single toss is p (and that of T is $q = 1 - p$), then, disregarding the logical difficulties, the probability that exactly m heads will show in n tosses is

$$\frac{n!}{m!(n-m)!} p^m q^{n-m}$$

Laplace (following earlier work of De Moivre) now proved that the probability that the number of heads in n trials will be between

$$np + \alpha\sqrt{2pqn} \qquad \text{and} \qquad np + \beta\sqrt{2pqn}$$

(α and β are given and fixed); will, as n gets larger and larger, be approximated better and better by the integral

$$\frac{1}{\sqrt{2\pi}} \int_\alpha^\beta e^{-\frac{x^2}{2}} dx$$

i.e., by the area under the curve

$$y = \frac{1}{\sqrt{2\pi}} e^{-\frac{x^2}{2}}$$

between $x = \alpha$ and $x = \beta$ (see fig. 8).

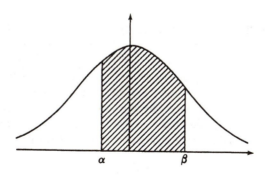

Fig. 8

In other words, as n approaches infinity, the limit of the probability that the number of heads shown will be between $np + \alpha \sqrt{2pqn}$ and $np + \beta \sqrt{2pqn}$ is the integral

$$\frac{1}{\sqrt{2\pi}} \int_{\alpha}^{\beta} e^{-\frac{x^2}{2}} dx$$

Why should this in any way be considered remarkable? Stripped of probabilistic terminology it becomes a rather special statement concerning the binomial coefficients $C(n;m)$; mathematical literature abounds in statements concerning such coefficients.

To appreciate the importance of Laplace's theorem (rightly, the De Moivre-Laplace Theorem) one must look outside mathematics.

First of all, not only is the theorem in *qualitative* agreement with one's intuition but it also makes the intuition more precise in a specific *quantitative* way. "Probability is common sense made precise" is the way Laplace himself put it.

If we toss an "honest" coin ($p = q = \frac{1}{2}$) 10,000 times, our intuition will lead us to expect about 5,000 heads. The De Moivre-Laplace Theorem tells us that with probability about 0.99998 we could expect the number of heads to lie between 4850 and 5150 and with probability about 0.8427 to lie between 4950 and 5050.

Secondly, the curve

$$y = \frac{1}{\sqrt{2\pi}} e^{-x^2/2}$$

or, somewhat more generally,

$$y = \frac{1}{\sigma\sqrt{2\pi}} e^{-(x - m)^2/2\sigma^2}$$

has been repeatedly encountered in empirical contexts. It is often called the normal curve or the curve of the normal distribution. That there was a *mathematical model* which led to this curve was certainly highly suggestive.

The history of probability theory provides us with an excellent example of the opposition between "pure" and "applied" motivations in mathematics. To the purist the De Moivre-Laplace Theorem was, at best, a contribution to the large body of specialized knowledge concerning binomial coefficients. The original proof was based on the famed asymptotic formula of Stirling

$$n! \sim \left(\frac{n}{e}\right)^n \sqrt{2\pi n}$$

The symbol $\sim$ means that the limit relation

$$\lim_{n \to \infty} n! / \left(\frac{n}{e}\right)^n \sqrt{2\pi n} = 1$$

holds.

Thus the purist could argue that whatever "depth" can be attributed to the theorem is there solely by the grace of Stirling. Finally, he would reject the probabilistic interpretation on grounds of logical insufficiency and remain unmoved by the argument that the De Moivre-Laplace Theorem is, without doubt, an important step toward analyzing the vast world of the phenomena of chance.

Because of this rigid attitude, probability theory all but disappeared as a mathematical discipline until its spectacular successes in physics revived interest in it early in the 20th century.

In retrospect, the logical difficulties of Laplace's theory proved to be minor, yet attempts at clarification of the foundations of probability theory had a distinctly beneficial effect on the subject.

The contemporary view is quite simple:

From the set Ω of all possible outcomes (called "sample space") a collection of subsets (called "elementary events") is chosen whose probabilities are assumed to be given once and for all. One then tries to calculate probabilities of more complicated events by the use of two axioms:

1. *Axiom of additivity:* If E_1, E_2 are mutually exclusive events (*i.e.*, the corresponding subsets in the sample space have no elements in common) then the probability (Prob.) of the event $\{E_1 \text{ or } E_2 \text{ or } \ldots\}$ is the sum of the probabilities of the constituent events; of course, provided the constituent events *can be assigned probabilities*.

Symbolically,

$$\text{Prob. } \{E_1 \text{ or } E_2 \text{ or } \ldots\} = \text{Prob. } \{E_1\} + \text{Prob. } \{E_2\} + \cdots$$

2. *Axiom of complementarity:* If an event E can be assigned a probability then the event "not E" also can be assigned a probability.

Finally the whole sample space is assigned (by convention) probability 1

$$\text{Prob. } \{\Omega\} = 1$$

so that, *e.g.*, (using both axioms)

$$\text{Prob. } \{\text{not } E\} = 1 - \text{Prob. } \{E\}$$

provided Prob. $\{E\}$ is defined.

Why these axioms? What is usually required of axioms is that they should codify intuitive assumptions and that they be directly *verifiable* in a variety of simple situations.

The axioms above clearly hold in all situations to which Laplace's definition is unambiguously applicable; they are also in accord with almost every intuition one has about probabilities.

An important exception is encountered in quantum mechanics. Let a source S of mono-energetic electrons be placed behind a screen (a) in which there are two small holes A and B (see fig. 9). The arrival of electrons at another screen (b) can be tested by appropriate detectors. If hole B is closed, electrons can go only through A and if they had behaved purely classically they would all arrive very near the point A′ at which the straight line connecting S with the centre of the hole A hits the screen (b). Actually the places of arrival are governed by "chance" and the best one can do is to associate with each region R on the screen (b) a probability $P_A(R)$ that an electron arriving at (b) will hit it inside R. Similarly $P_B(R)$ is the probability that an electron emitted at S will arrive at (b) in R if hole A is closed.

If both holes A and B are open one could argue that the probability $P_{A \text{ or } B}(R)$ should be the sum $P_A(R) + P_B(R)$. It is well established experimentally that this is not the case and

$$P_{A \text{ or } B}(R) \neq P_A(R) + P_B(R)$$

Thus the seemingly obvious statement that an electron reaching (b) *has* to go through either A or B is untenable. A subtle analysis shows that an *experiment* designed to determine through which hole an electron actually does go *interferes* so strongly with the motion of the electron that the axiom of additivity is restored!

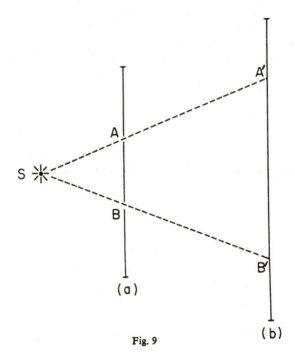

Fig. 9

The axioms of additivity and complementarity are also too general and too all-embracing to stand alone as a foundation for a theory as rich and as fruitful as probability theory. Yet, as will be seen from the next section, these two axioms are far from being without an impressive mathematical content.

Selection of "elementary" events and decision on what probabilities to assign them lie at the heart of the subject. Here nonmathematical considerations come into play and we must rely upon the empirical world to guide us toward promising areas of exploration.

Let us return to the experiment of tossing a coin n times. In attempting to construct any kind of a realistic and useful theory we must first consider two entirely different questions:

(1) What kind of coin is being tossed?

and

(2) What is the tossing mechanism?

The first question has something to do with the way the coin is made; the second with whether successive tosses are correlated (and if so, how).

To appreciate these points more fully consider the statistical structure of English. Letters in English texts appear with certain frequencies that vary remarkably little from text to text. Thus e empirically has been found to account for about 13.05% of all letters, t for 9.02%, and a for 6.81%. We could now imagine a 26-faced die with its faces marked by the letters of the alphabet and "loaded" in such a way that the probabilities of various faces are equal to frequencies with which their letters are used in the language. We could even add a 27th face marked "space" and reload the die appropriately. If we now toss this die (say 10,000 times) we will obtain a text in which letters and "space" will have nearly the same frequencies as in English, but will not look anything like it unless care is taken to *correlate* the tosses in a specific way. For we know that h is more likely to follow t than any other letter, n is more likely to be the last letter of a word (*i.e.*, to precede "space") than any other, and so on. If these correlations are taken into account, the resulting text will look much more like English, and by the time the frequencies of trigrams (successions of three consecutive letters) are adjusted, the text may actually fool you at first brief glance.

Experiments of this sort were conceived and performed by Claude Shannon in connection with his pioneering and beautiful work on information theory.

Returning to our coin, the simplest assumptions are that it is "honest"; H and T are to be assigned probability $\frac{1}{2}$ and the tosses are independent.

Since the notion of *independence* is central to probability theory we must discuss it in some detail.

Events E and F are independent in the ordinary sense of the word if the occurrence of one has no influence on the occurrence of the other.

In such a case it should clearly be possible to calculate the probability of the composite event "E and F" if one only knows the probabilities of the constituent events E and F.

In other words, whenever E and F are independent, there should be a *rule* that would make it possible to calculate Prob. $\{E$ and $F\}$ provided only that one knows Prob. $\{E\}$ and Prob. $\{F\}$. Moreover, this rule should be *universal;* it should be applicable to every pair of independent events.

Such a rule takes on the form of a function $f(x, y)$ of two variables x, y, and we can summarize by saying that whenever E and F are independent we have

$$\text{Prob. } \{E \text{ and } F\} = f(\text{Prob. } \{E\}, \text{Prob. } \{F\})$$

Let us now consider the following experiment. Imagine a coin that can be "loaded" in any way we wish (*i.e.*, we can make the probability p of H any number between 0 and 1) and a four-faced die that can be "loaded" to suit our purposes also. The faces of the die will be marked 1,2,3,4 and their respective probabilities will be denoted p_1, p_2, p_3, p_4; each p_i is nonnegative and $p_1 + p_2 + p_3 + p_4 = 1$. We must now assume that whatever independence means, it should be possible to toss the coin and the die independently. If this is done and we consider (*e.g.*) the event "H and (1 or 2)" then on the one hand

$$\text{Prob. } \{H \text{ and } (1 \text{ or } 2)\} = f(p, p_1 + p_2)$$

while on the other hand, since the event "H and (1 or 2)" is equivalent to the event "(H and 1) or (H and 2)," we also have

$$\text{Prob. } \{H \text{ and } (1 \text{ or } 2)\} = \text{Prob. } \{H \text{ and } 1\} + \text{Prob. } \{H \text{ and } 2\}$$
$$= f(p, p_1) + f(p, p_2)$$

Note that we have used the axiom of additivity repeatedly. Thus

$$f(p, p_1 + p_2) = f(p, p_1) + f(p, p_2)$$

for all p, p_1, p_2 restricted only by the inequalities

$$0 \leqslant p \leqslant 1, \qquad 0 \leqslant p_1, \qquad 0 \leqslant p_2, \qquad p_1 + p_2 \leqslant 1$$

If one assumes, as seems proper, that f depends continuously on its variables, it follows that $f(x,y) = xy$ and hence the probability of a joint occurrence of independent events should be the *product* of the individual probabilities.

This discussion (which we owe to H. Steinhaus) is an excellent illustration of the kind of informal (one might say "behind the scenes") argument that precedes a formal definition. The argument is of the sort that says in effect: "We do not really know what independence is, but whatever it is, if it is to make sense, it must have the following properties. . . ." Having drawn from these properties appropriate consequences (*e.g.*, that $f(x,y) = xy$ in the above discussion), a mathematician is ready to tighten things logically and to propose a *formal definition.*

Technically, the two events E and F (or any finite number for that matter) are said *to be independent* if the *rule* of *multiplication of probabilities* is applicable; *i.e.*,

$$\text{Prob. } \{E \text{ and } F\} = \text{Prob. } \{E\} \cdot \text{Prob. } \{F\}$$

There is another way of justifying the rule of multiplication of probabilities for independent events. It is based on the notion of frequency. Suppose that in n trials event E occurred $n(E)$ times, event F occurred $n(F)$ times, and the two occurred simultaneously $n(E \text{ and } F)$ times.

Intuitively one clings to the notion that the frequency with which an event occurs in a long series of trials should in some sense approximate its probability.[7]

Now

$$\frac{n(E \text{ and } F)}{n} = \frac{n(E \text{ and } F)}{n(F)} \frac{n(F)}{n}$$

and we see that the frequency with which E and F occur simultaneously in n trials is the product of the frequency with which E occurs in the $n(F)$ trials in which F occurs, times the frequency with which F occurs.

One might now argue that, if E and F are independent, the knowledge that F occurred should be of no help in predicting the occurrence or nonoccurrence of E. Consequently, the frequency with which E occurs during the trials in which F occurred should be approximately the same as the overall frequency of E in n trials.

In other words we might expect approximately that

$$\frac{n(E \text{ and } F)}{n(F)} = \frac{n(E)}{n}$$

and consequently

$$\frac{n(E \text{ and } F)}{n} = \frac{n(E)}{n} \frac{n(F)}{n}$$

which is strongly reminiscent of

$$\text{Prob. } \{E \text{ and } F\} = \text{Prob. } \{E\} \cdot \text{Prob. } \{F\}$$

This "justification" is only heuristic but it gets us back to the rule of multiplication of probabilities, and in a context that is wholly different from the previous one.[8]

Having agreed to interpret independence to mean that the rule of multiplication of probabilities is applicable, we go back to the loaded coin of Laplace.

---

[7] Physicists (somewhat uncritically) actually identify probabilities with frequencies. There are all sorts of difficulties in such an approach since one must take limits as the number of trials becomes larger and larger. Richard von Mises tried to axiomatize probability theory on the basis of frequencies in certain *infinite* sequences of trials that he called *collectives*. Some early logical difficulties in this approach were repaired by Abraham Wald, but the approach did not achieve widespread acceptance among mathematicians.

[8] While in a formal sense we have proved nothing, we are reinforced in our feeling that we are on the right track because things somehow "hang together." In theoretical physics "hanging together" is an important (and often the sole) guide to truth. In mathematics there is always the possibility that some hidden inconsistency or a subtle false premise may have caused one to think that all is well while in reality some gigantic paradox is at the bottom of it all. One must indeed believe with Einstein that *raffiniert ist der Herr Gott aber boshaft ist er nicht* ("The Lord God is subtle, but he is not malicious.")

If the tosses are assumed to be independent, the probability associated with a *specified* word like

$$HHTT \ldots TH$$

containing m H's (and $n - m$ T's) is $p^m q^{n-m}$. Since there are $C(n;m)$ such words (using the axiom of additivity) the probability of the event that exactly m out of n *independent* tosses will be heads is $C(n;m)p^m q^{n-m}$.

We have arrived at the same formula but with no reference to "equally likely"; instead we have related it to the concept of independence.

That this constitutes more than a mere translation became apparent when analysis showed that the normal curve

$$y = \frac{1}{\sqrt{2\pi}} e^{-x^2/2}$$

is primarily the result of the independence of trials and not of the magic of Stirling's formula.

Consider, for example, a sequence of differently loaded coins that are being tossed independently (sometimes called the Poisson scheme). Let p_k and $q_k = 1 - p_k$ be respectively the probabilities of H and T of the kth coin.

There is now no compact and simple formula for the probability that in n independent tosses exactly m heads will show.

But using an adaptation of the method of generating functions discussed in Section 6 earlier one can still prove the following generalizations of the De Moivre-Laplace Theorem.

The probability that the number of heads in n tosses lies between

$$(p_1 + \cdots + p_n) + \alpha\sqrt{p_1 q_1 + \cdots + p_n q_n}$$

and

$$(p_1 + \cdots + p_n) + \beta\sqrt{p_1 q_1 + \cdots + p_n q_n}$$

(α and β given and fixed) in the limit as $n \to \infty$ becomes the integral

$$\frac{1}{\sqrt{2\pi}} \int_\alpha^\beta e^{-x^2/2} dx$$

provided only the infinite series $p_1 q_1 + p_2 q_2 + \cdots$ diverges.[9]

Except then for a mild proviso concerning divergence of a certain series the normal law as expressed by the curve

$$y = \frac{1}{\sqrt{2\pi}} e^{-x^2/2}$$

---

[9] It also should be noted that this condition is quite natural. If the series in question were convergent, $p_n q_n$ would have to be very small for large n; *i.e.*, the high-numbered coins would be so heavily loaded in favour of either H or T that an unbearable strain would be put on the "laws of chance."

assumes an aura of universality at least in the realm of independent trials and events.

The extent to which the normal law is universal in this realm was determined during the 1920s and 1930s. The findings are too technical to be presented here, but in the process much was developed that proved valuable in other parts of mathematics and science. This is a tribute to Laplace's judgment of the importance of probability theory.

Today the De Moivre-Laplace Theorem and its extension to Poisson schemes are only special corollaries of the very general Central Limit Theorem. But, as has been the case with other great theorems, they contained most of the seeds of the generality that ultimately engulfed and subsumed them.

> The realization that independence is mainly responsible for the normal law makes it possible to apply theorems like that of De Moivre-Laplace to situations that are far removed from chance phenomena.
>
> As an example let us consider the positive integers 1, 2, 3, 4, ... and the prime numbers 2, 3, 5, 7, 11, ... which (since we have used p_1, p_2, ... to denote probabilities in this section) will be denoted P_1, P_2, ... (thus, $P_1 = 2$, $P_2 = 3$, ...).
>
> Let E be a subset of integers and let $K_n(E)$ denote the number of elements of E among the first n positive integers 1, 2, 3, ..., n. If as n approaches ∞ the limit
>
> $$D\{E\} = \lim_{n \to \infty} \frac{K_n(E)}{n}$$
>
> exists [10] we call it the *density* of E.
>
> Let E_i be the set of integers divisible by the ith prime P_i.
>
> It is almost immediate that
>
> $$D\{E_i\} = \frac{1}{P_i}$$
>
> Consider now the set $E_1 \cap E_2 \cap \ldots \cap E_r$ of integers divisible by the first r primes P_1, P_2, ... P_r. (The symbol $\cap$ stands for "intersection"; *i.e.*, $E_1 \cap E_2 \cap \ldots \cap E_r$ is the set of elements common to E_1, E_2, ... E_r.) Again it is almost immediate that
>
> $$D\{E_1 \cap E_2 \cap \ldots \cap E_r\} = \frac{1}{P_1 P_2 \ldots P_r} = D\{E_1\}\, D\{E_2\} \ldots D\{E_r\}$$
>
> and the analogy with the rule of multiplication of probabilities emerges.
>
> Using this analogy and stretching the application of the Poisson extension of the De Moivre-Laplace Theorem a bit one is led to suspect that the density of the set of integers n for which the number of prime divisors $\nu(n)$ lies between $\log \log n + \alpha\sqrt{\log \log n}$ and $\log \log n + \beta\sqrt{\log \log n}$ is the integral

---

[10] It should be noted that often the existence of a limit is by far the deeper part of a theorem while the actual numerical determination of the limit is relatively easy. For example, the deep part of the prime-number theorem mentioned in Section 1 is that the limit

$$\lim_{n \to \infty} \frac{\pi(n)}{n/\log n}$$

exists; once this is established the fact that it is equal to 1 follows quite simply.

$$\frac{1}{\sqrt{2\pi}} \int_{\alpha}^{\beta} e^{-x^2/2} \, dx$$

Examples of $\nu(n)$ include: $\nu(20) = \nu(2^2 \cdot 5) = 2, \nu(90) = \nu(2 \cdot 3^2 \cdot 5) = 3, \nu(13) = 1.$

Symbolically,

$$D\{\log \log n + \alpha\sqrt{\log \log n} < \nu(n) < \log \log n + \beta\sqrt{\log \log n}\} = \frac{1}{\sqrt{2\pi}} \int_{\alpha}^{\beta} e^{-x^2/2} \, dx$$

The suspicion has been confirmed. We have the theorem that the number of prime divisors are indeed distributed according to the normal law.

Here is an example of how concepts and methods find significant uses in an area far removed from the one whose conditions inspired their creation.

And here is the normal law, so closely associated in popular thinking with randomness and chance, making its appearance in number theory, the tightest and least "random" branch of pure mathematics.

10. *Measure*

The problem of measuring regions in the plane or in 3-space with the view of assigning them numbers representing areas or volumes can be traced to the very beginning of mathematics. The Greeks developed a systematic theory of area and volume for polygons and polyhedra. Integral calculus (whose origin can be traced to Archimedes) extended the theory to deal with a large class of regions bounded by curves or curved surfaces.

But as mathematics continued to develop a need arose to assign measures to an even wider class of sets, and this led to the development of a general measure theory by Borel and Lebesgue. Even this extension is not complete, and there are sets that cannot be assigned a Lebesgue measure (*i.e.*, they are nonmeasurable). Construction of nonmeasurable sets involves the use of the celebrated axiom of choice: given a collection C of disjoint sets one can choose one element out of each set of the collection and combine the selected elements to form a set Z. This innocent-sounding axiom has many consequences that may seem strange and paradoxical.

Measure theory has important applications in more advanced parts of probability theory as discussed briefly in Section 11.

Many mathematical ideas, that now appear to be algebraic or analytic, have their origin in problems that originally were geometric in character. Such is the case with the notion of measure and measuring.

Consideration of the *length* of a straight or curved segment, the value of an area, or volume of a region, probably came with the earliest attempts to use numbers for more than mere counting of discrete objects. The origins of these notions go far back in time and are contemporary with the earliest mathematical attempts by the Babylonians, Egyptians, and the Greeks.

Euclid dealt mainly with areas of polygons and volumes of polyhedra. In this limited context two axioms were used:

1. If polygons (polyhedra) A and B are congruent, they have the same area (volume).

2. If a polygon (polyhedron) A is decomposable into a *finite* number of disjoint polygons (polyhedra) A_1, A_2, ..., A_n then the area (volume) of A is the sum of the areas (volumes) of the constituent pieces A_1, A_2, ..., A_n.[11]

If, in addition, a certain square and a certain cube (or an interval on a straight line) are chosen to have *unit area* and *unit volume* one can unambiguously assign numerical values to polygons and polyhedra that represent their areas and volumes in the chosen units.

Even the problem of determining the area of a circle required passing the safe bounds of the finite and introducing the genuinely infinitesimal operation of taking a limit. Among the ancients, Archimedes contributed most to determining areas and volumes of figures bounded by curves and curved surfaces; there is good evidence that he was fully aware of the subtlety of the concept of limit. Without doubt this was the beginning of the integral calculus, but not until the time of Newton and his followers did the calculation of areas and volumes become fully systematized.

Calculus allowed one to assign areas, volumes, and lengths only to relatively "tame" sets. It provided no machinery to deal (*e.g*) with sets like the set of all points (x,y) where both x and y are rational numbers lying between 0 and 1. But then such sets did not arise as long as the problematics of calculus mainly reflected the needs of physics and geometry.

In the latter part of the 19th century problems were emerging that led to a need to assign numerical measures to a much wider collection of sets than hitherto had been considered.

Growing preoccupation with problems of convergence and divergence focused attention on sets of convergence or divergence and on the problem of determining their "size." In fact, Cantor's set theory, which ultimately became the cornerstone of all of modern mathematics, originated in his interest in trigonometric series and their sets of convergence.

The problem of measure can be formulated quite simply.

One wants to be able to assign to a set A a *nonnegative* number $m(A)$, which will be called the *measure of A*, with the following properties:

1. If A_1, A_2, ... are disjoint sets that are *measurable, i.e.,* each A_i can be

---

[11] Strictly speaking, such a decomposition is impossible since two adjacent polygons (polyhedra) must have parts of their boundaries in common, or if polygons (polyhedra) are defined not to include their boundaries (*i.e.*, they are "open" polygons) they cannot provide a completely exhaustive decomposition.

assigned a measure $m(A_i)$, then their *union* $A_1 \cup A_2 \cup \ldots$ (*i.e.*, the set consisting of elements of A_1, A_2, ...) is also measurable. Moreover

$$m(A_1 \cup A_2 \cup \ldots) = m(A_1) + m(A_2) + \ldots$$

2. If A and B are measurable and A is contained in B ($A \subset B$), then $B - A$ (*i.e.*, the set of points in B that are *not* in A) is also measurable; which by property 1 implies that $m(B - A) = m(B) - m(A)$.

3. A certain set E (unit set) is assumed to have measure 1

$$m(E) = 1$$

4. If two sets are congruent their measures are equal (provided they are measurable).

In dealing with sets of points on a line, E is taken to be an interval; in the plane, E is a square; and in space, E is a cube. This choice is dictated by a desire to have measures assigned to "tame" sets agree with those assigned to them previously in geometry or calculus.

Can one significantly enlarge the class of sets to which measures can be assigned in accordance with the above properties or rules?

The answer is a resounding yes provided (and it is a crucial proviso) that in property 1 we allow *infinitely many A*'s.

If, following Euclid, one allows only a finite number of A's (in which case the measure is called *finitely additive*), one gains almost nothing and the enlargement of the class of "tame" sets is quite insignificant. For example, the set of points with rational coordinates within the unit square can be shown to be nonmeasurable; *i.e.*, it cannot be assigned a measure without running into a contradiction.

The situation changes dramatically if one considers *completely additive* measures; *i.e.*, if one allows infinitely many sets A in property 1 above. Now, the class of measurable sets has been enlarged enormously and essentially all sets used in classical mathematics and those defined in modern mathematics are measurable.

Completely additive measures were introduced early in this century by Émile Borel and Henri Lebesgue [12] to originate a most vigorous and fruitful line of inquiry in mathematics. Lebesgue's measure is, in fact, one of the most powerful tools of modern analysis.

But how general is this measure? Can one assign it to every set on the line? Vitali first showed that even the Lebesgue measure has its limitation; *i.e.*, that there are sets of points for which it cannot be defined.

It is easier to explain it on the circumference of a circle than on the line but the problem is essentially the same. (Two sets will be congruent on a circle if

---

[12] In addition to the four properties listed above Lebesgue also postulated a fifth one that a subset of a set of measure zero is measurable (and hence of measure zero).

they can be brought to coincide by a rotation of the circle.) We shall now define a set Z of points (angles) on the circle, for which no measure with the above properties is possible. Given a point x on the circle, consider with it all points that can be obtained from it by rotating the circle through angles that are rational fractions of 2π. Such angles form a countable set and we will call them $\alpha_1, \alpha_2 \ldots \alpha_n \ldots$ The set of *all* angles of the circle can be split into mutually disjoint classes C such that in every one of these classes the points differ by a rational value. These classes have no elements in common. What we do now is to pick from each of these classes exactly one point. Call the totality of all selected points Z. If we now rotate the set Z by angles $\alpha_1, \alpha_2 \ldots$ we obtain new sets, *all disjoint among themselves*. Their totality is the set of all points on the circle. These sets, in addition, are countable in number and are mutually congruent, since they originate from each other by successive rotations: $\alpha_1, \alpha_2 \ldots$ We now have obtained a countable collection of sets: $Z, Z_1 \ldots Z_n \ldots$, etc., with the following properties: they are mutually disjoint, congruent to each other, and their union is the whole circle which we may choose to be our "unit set" E where $m(E) = 1$. What could the measure of Z be? If it should be 0 we get a violation of our additivity postulation because the measures of all these sets would be 0 (since they are congruent) but their countable sum must have measure 1, since $m(E) = 1$. This violates the postulate of additivity (property 1 above). Should the measure of Z be positive, we would get a sum of infinitely many equal positive numbers, which again violates additivity. (The measures should add up to 1 and not to infinity.)

We have therefore exhibited a set for which no measure can be defined. Here is again a proof of *impossibility* showing that the process of generalization, as elsewhere in mathematics, must go on forever. However in defining the set Z of Vitali we made use of a highly nonconstructive device. In fact, we were asked to *pick* one element out of each class C. But how? The classes C are not defined explicitly enough to permit an explicit rule of choosing elements from them. And yet we feel that we should be allowed to pick an element out of each class and combine these into a new set even if we are unable to give a concrete prescription for performing this task.

A way out of this dilemma was suggested at the beginning of the 20th century by Ernest Zermelo. Zermelo proposed to legitimize "constructions" like that of the set Z above by adopting the following general axiom:

Given a collection C of disjoint sets one can form a set Z by *choosing* one element out of each set of the collection.

This axiom became known as the *axiom of choice*. Since its birth it has been a point of debate and controversy, for many of its consequences appeared strange and paradoxical.

For example, Banach and Tarski proved that given two spheres S_1 and S_2 of *different* radii it is possible to divide each into the same finite number of disjoint sets, all congruent to each other! In other words there are decompositions

$$S_1 = A_1 + A_2 + \ldots A_n$$
$$S_2 = B_1 + B_2 + \ldots B_n$$

where all the A_i and all the B_i among themselves are pairwise disjoint while A_i is congruent for all $i = 1, 2, \ldots$ to B_i. It is impossible to define measures for these sets since, by putting them (disjointly) together in one fashion, we obtain a larger sphere while, by making a different spatial arrangement, we get a smaller one! By the way, such a decomposition is not possible in the plane. In fact, Banach proved that one can find a *finitely* additive measure for all subsets of the plane with the property that congruent sets have equal measure.

Attempts to generalize the notion of measure were made from necessity. As already mentioned, more and more general sets were considered by mathematicians. In the study of trigonometric series, for example, one could formulate theorems that were valid for all real numbers *except* for those belonging to a specific set. One wanted to state in a rigorously defined way that the set of these exceptional points is in some sense small or negligible. One could "neglect" merely countable sets as small in the noncountable continuum of all points but in most cases the exceptional sets turned out to be noncountable, though still of Lebesgue *measure* 0. In the theory of probability one has many statements that are valid "with probability one" (or "almost surely"). This simply means that they hold for "almost all" points of an appropriate set; *i.e.*, for all points except for a set of measure 0. In statistical mechanics one has important theorems that assert properties of dynamic systems that are valid only for *almost all* initial conditions.

One final remark:

The notion or concept of measure is surely close to the most primitive intuition. The axiom of choice, that simply permits one to consider a new set Z obtained by putting together an element from each set of a family of disjoint sets, sounds so obvious as to be nearly trivial. And yet it leads to the Banach-Tarski paradox!

One can see why a critical examination of the logical foundation of set theory was absolutely necessary and why the question of existence of mathematical constructs became a serious problem.

If to exist is to be merely free from contradiction as Poincaré decreed, we have no choice but to learn to live with unpleasant things like nonmeasurable sets or Banach-Tarski decompositions.

11. *Probability Revisited*

It could hardly have escaped the reader's notice that the axioms of additivity and complementarity that form the general basis for probability theory are *identi-*

cal with the axioms of measure. Thus probabilities are measures, and probability theory is a part of measure theory.

The recognition that this is so, though rather trivial in itself, had serious and even profound implications.

First of all it made probability theory "respectable" by supplying a rigorous framework. Secondly, and far more importantly, it greatly increased its scope by making possible the posing and solution of a new complex of problems.

In Section 8 we did not specify whether the axiom of additivity should hold for a finite collection of events. It is now universally accepted that one should allow infinite collections, thus making the measure *countably additive*.

The main reason for this is to be able to deal with problems that involve infinite sequences of trials.

The necessity for countable additivity is present even in very elementary situations. If, for example, two persons A and B alternate in tossing two coins (first A, then B, then A, and so on) it may be of interest to find the probability that A will be the first to toss heads. This can happen either on the first toss, or on the third (the first two being T's) or on the fifth (the first four being T's), and so on. The event that A will toss the first H is thus decomposed in a rational way into an infinite number of mutually exclusive events. If the coins are "honest" and the tosses independent, the probabilities of constituent events are

$$\frac{1}{2}, \frac{1}{2^3}, \frac{1}{2^5} + \cdots$$

and the desired answer is

$$\frac{1}{2} + \frac{1}{2^3} + \frac{1}{2^5} + \cdots = \frac{2}{3}$$

provided the axiom of additivity can be applied.

Let us now consider the following problem. Suppose we keep tossing an "honest" coin and suppose that the tosses are independent. What is the frequency with which H will show?

Our intuition will tell us that the answer should be $\frac{1}{2}$. But what does it mean? Surely, even in a very long series of tosses, it would be foolish to expect *exactly* half to be H's and exactly half to be T's. It should be clear that what one is looking for is a statement that holds only *in the limit as the number of tosses becomes infinite*.

One such statement is already available within Laplace's framework and it is a consequence of the De Moivre-Laplace Theorem discussed in Section 8.

If ϵ is a positive number let $p_n(\epsilon)$ denote the probability that the frequency of H's in n trials will differ from $\frac{1}{2}$ by more than ϵ. It then follows that $p_n(\epsilon)$ approaches 0 as n approaches infinity

$$\lim_{n \to \infty} p_n(\epsilon) = 0$$

This theorem (first proved by Bernoulli and known as the "weak law of large numbers") states that no matter what (positive) ϵ is chosen, we can make the probability that the deviation between the frequency and the probability exceeds ϵ arbitrarily small by taking a sufficiently large number of trials. The measure-

the potential is equal to unity. Since the calculation of such a potential can be reduced by classical methods to solving a specific differential equation, we establish in this way a significant link between classical analysis and measure theory. Much work along these lines is being done and the subject is still developing as mathematics faces the 1970s.

12.   *Groups and Transformations*

One of the most important, fruitful, and all-embracing concepts in mathematics is that of a *group*. It is convenient to introduce this concept in conjunction with that of a *transformation* of a set into itself.

Let S be a set; a transformation f of S into itself is simply a way of assigning to each element p of S a *unique* element $f(p)$ of S, where $f(p)$ is called the *image of p under f*.

If $f(p) = p$ for *every p*, then f is the *identity transformation;* and if $f(p)$ is one-to-one, *i.e.*, if $p \neq q$ then $f(p) \neq f(q)$, one can define the *inverse transformation* f^{-1} as follows:

$$f(p) = q \text{ then } f^{-1}(q) = p$$

In other words the image of q under f^{-1} is that (unique) element whose image under f is q.

Given two transformations f and g of S into itself, one can define a new transformation fg as the transformation that results from first applying g and then f. In other words the image of p under fg is the image under f of the image under g of p. Symbolically,

$$fg(p) = f(g(p))$$

One can also consider gf defined symbolically by the formula

$$gf(p) = g(f(p))$$

and in general gf is *not* the same as fg.

However the operation of *composing* transformations (*e.g.*, of forming fg or gf) is *associative; i.e.*,

$$(fg)h = f(gh)$$

This can be verified by considering what each of the two transformations, $(fg)h$ and $f(gh)$, does to a representative element p of S and by discovering that the final outcomes are the same.

A (finite or infinite) collection G of transformations is said to form a group if:

1. Whenever f and g belong to the collection, then so does their composition fg (and, of course, gf).
2. The identity transformation belongs to the collection.
3. Whenever a transformation f belongs to the collection, so does its inverse f^{-1}.

If the set S is finite, the one-to-one transformations of S into itself consist merely in changing the *order* of elements. Such transformations are called *permutations*. If S consists of n objects, there are $n! = 1 \times 2 \times 3 \times \ldots \times n$ different permutations. Denoting the objects $1, 2, 3, \ldots, n$ we associate with a permutation f the symbol

$$f = \begin{pmatrix} 1, & 2, & 3, & \ldots, & n \\ f(1), & f(2), & f(3), & \ldots, & f(n) \end{pmatrix}$$

where $f(k)$ is the image of k under the permutation f. It should be clear that $f(i)$ is different from $f(j)$ whenever i is different from j. (This is another way of stating that f is a one-to-one transformation of S into itself.)

Here, for example, are all 6 ($= 3! = 1 \times 2 \times 3$) permutations of a set of 3 elements.

$$f_0 = \begin{pmatrix} 1\ 2\ 3 \\ 1\ 2\ 3 \end{pmatrix} \qquad f_1 = \begin{pmatrix} 1\ 2\ 3 \\ 1\ 3\ 2 \end{pmatrix} \qquad f_2 = \begin{pmatrix} 1\ 2\ 3 \\ 2\ 1\ 3 \end{pmatrix}$$

$$f_3 = \begin{pmatrix} 1\ 2\ 3 \\ 2\ 3\ 1 \end{pmatrix} \qquad f_4 = \begin{pmatrix} 1\ 2\ 3 \\ 3\ 1\ 2 \end{pmatrix} \qquad f_5 = \begin{pmatrix} 1\ 2\ 3 \\ 3\ 2\ 1 \end{pmatrix}$$

We note that f_0 is the identity transformation, $f_1^2 = f_2^2 = f_5^2 = f_0$ (*i.e.*, $f_1^{-1} = f_1$, $f_2^{-1} = f_2$, $f_5^{-1} = f_5$), $f_3 f_4 = f_0$, $f_1 f_4 = f_2$, $f_4 f_1 = f_5$, and so on. In fact we have the following "multiplication table" for the six permutations $f_0, f_1, \ldots, f_5$.

	f_0	f_1	f_2	f_3	f_4	f_5
f_0	f_0	f_1	f_2	f_3	f_4	f_5
f_1	f_1	f_0	f_4	f_5	f_2	f_3
f_2	f_2	f_3	f_0	f_1	f_5	f_4
f_3	f_3	f_2	f_5	f_4	f_0	f_1
f_4	f_4	f_5	f_1	f_0	f_3	f_2
f_5	f_5	f_4	f_3	f_2	f_1	f_0

Groups of permutations were first introduced by Abel and Galois in connection with their celebrated studies on solvability of algebraic equations in terms of radicals.

Some of the underlying ideas are of such fundamental importance and they exemplify so well the spirit of algebra that we shall give a brief and elementary account of them.

Consider the cubic equation

$$x^3 + ax^2 + bx + c = 0$$

It has three roots x_1, x_2, x_3 that are in general distinct. In terms of these roots the coefficients a, b, c are expressed by the well-known elementary formulas

$$a = -(x_1 + x_2 + x_3)$$
$$b = x_1x_2 + x_2x_3 + x_3x_1$$
$$c = -x_1x_2x_3$$

A function $F(x_1, x_2, x_3)$ is said to be *symmetric* if its value remains unchanged when its arguments x_1, x_2, x_3 are subjected to a permutation. We thus see that the coefficients of a cubic equation are symmetric functions of its roots. In general, a function will not be symmetric; it is proper to ask: Which permutations leave it unchanged?

For example, the function

$$\Delta \equiv (x_1 - x_2)(x_1 - x_3)(x_2 - x_3)$$

will remain unchanged under permutations f_0, f_3, and f_4 but will undergo a change in sign under the remaining permutations.

The function

$$\Phi \equiv a_1x_1 + a_2x_2 + a_3x_3$$

will in general be changed by every f except f_0.

Let us consider the permutations that leave a given function unchanged. They form a *subgroup; i.e.,* a subset of the group that is itself a group. (In other words, a subgroup will contain the identity permutation f_0, and whenever it contains f_i and f_j it contains f_if_j; it also contains the inverse of every one of its elements.)

Suppose now that a polynomial function Ψ is invariant under the subgroup (f_0, f_3, f_4) which leaves Δ unchanged. We shall prove that Ψ must be of the form

$$\Psi = A(x_1, x_2, x_3) + B(x_1, x_2, x_3)\Delta$$

where A and B are *symmetric* polynomial functions.

Our assumption is that for *all* x_1, x_2, x_3

$$\Psi(x_1, x_2, x_3) = \Psi(x_2, x_3, x_1) = \Psi(x_3, x_1, x_2)$$

so that, in particular, transposing x_1 and x_2 we obtain

$$\Psi(x_2, x_1, x_3) = \Psi(x_1, x_3, x_2) = \Psi(x_3, x_2, x_1)$$

We may as well assume that (in general)

$$\Psi(x_2, x_1, x_3) \neq \Psi(x_1, x_2, x_3)$$

since otherwise Ψ would be symmetric, making our statement trivial. (We would simply have $B = 0$.)

The polynomial $\Psi(x_1, x_2, x_3) - \Psi(x_2, x_1, x_3)$ is thus not identically equal to zero but assumes the value 0 when $x_2 = x_1$; consequently $\Psi(x_1, x_2, x_3) -$

$\Psi(x_2, x_1, x_3)$ is *divisible* by $x_2 - x_1$. By a similar argument we show that it is also divisible by $x_1 - x_3$ and by $x_2 - x_3$ and hence by Δ.

Thus we have

$$\Psi(x_1, x_2, x_3) - \Psi(x_2, x_1, x_3) = B(x_1, x_2, x_3)\Delta$$

and since

$$\Psi(x_2, x_1, x_3) = \Psi(x_1, x_3, x_2) = \Psi(x_3, x_2, x_1)$$

also

$$\Psi(x_1, x_2, x_3) - \Psi(x_1, x_3, x_2) = B(x_1, x_2, x_3)\Delta$$

and

$$\Psi(x_1, x_2, x_3) - \Psi(x_3, x_2, x_1) = B(x_1, x_2, x_3)\Delta$$

It follows immediately that B remains unchanged by the permutations f_1, f_2, and f_5; since $f_4 = f_1 f_2$ and $f_3 = f_2 f_1$ one concludes that B remains unchanged by *all* the permutations and is therefore symmetric.

Similarly one proves that $\Psi(x_1, x_2, x_3) - B\Delta$ is symmetric, and denoting it A we have our assertion that

$$\Psi = A + B\Delta$$

We have gone into such detail in deriving this rather special result because it is based on the immensely powerful and fruitful idea that much can be learned about the *structure* of certain objects by merely studying their *behaviour under the action of certain groups*.

In physics, for example, by studying the group of transformations that leave invariant the forces that hold atoms or molecules together, one can derive far-reaching results concerning the behaviour of their spectra (*e.g.*, one can explain so-called selection rules). As in the vast new world of elementary particles, even if one does not know the basic interactions, one can still gain much insight by postulating a fundamental symmetry with respect to a certain group (the much-heralded SU_3 in this case).

In the context of algebraic equations let us show how our considerations lead to the proof that a cubic is solvable in terms of radicals.

Let

$$\omega = -\frac{1}{2} + \frac{\sqrt{3}}{2}i$$

so that

$$\omega^2 = -\frac{1}{2} - \frac{\sqrt{3}}{2}i$$

and

$$\omega^3 = 1$$

Consider $\Psi = (x_1 + \omega x_2 + \omega^2 x_3)^3$ and note that it remains unchanged by f_3 and f_4. In fact, applying f_3 to Ψ we obtain $(x_2 + \omega x_3 + \omega^2 x_1)^3 = \omega^3(x_2 + \omega x_3 + \omega^2 x_1)^3 = (\omega x_2 + \omega^2 x_3 + x_1)^3$.

Thus $(x_1 + \omega x_2 + \omega^2 x_3)^3 = A(x_1,x_2,x_3) + B(x_1,x_2,x_3)\Delta$ and it can be shown that A, B, and Δ^2 are expressible as polynomials in a,b,c. Consequently

$$x_1 + \omega x_2 + \omega^2 x_3 = \sqrt[3]{A + B\Delta}$$

One also observes that $(x_1 + \omega^2 x_2 + \omega x_3)^3$ is unchanged by f_3 and f_4, and one obtains

$$x_1 + \omega^2 x_2 + \omega x_3 = \sqrt[3]{\bar{A} + \bar{B}\Delta}$$

with $\bar{A}$ and $\bar{B}$ again expressible as polynomials in the coefficients. Since as noted above

$$x_1 + x_2 + x_3 = -a$$

we can solve for x_1, x_2, and x_3 and obtain formulas in terms of cube and square roots. It should be recalled that Δ^2 is a polynomial in the coefficients so that Δ involves a square root.

If one explicitly executes the above steps, one is led to the famous formulas of Cardano discovered in the 16th century. At the time of their discovery the connection between algebraic equations and groups of permutations was not known. The discovery by Galois and Abel of this connection explained what made the method of solution "tick" and suggested a possible extension to quartics and equations of higher degree.

Every permutation of n objects can be expressed as a product of transpositions (*i.e.*, permutations that transpose two objects leaving the others alone). Though such a decomposition is not unique in general, a permutation *cannot* be decomposed into an even and into an odd number of transpositions. This leads to a natural division of permutations into *even* (which are composed of an even number of transpositions) and *odd* (which are composed of an odd number of transpositions). All even permutations (half the permutations of the whole group) form what is called the *alternating subgroup*. In the case $n = 3$ the permutations f_0, f_3, f_4 form the alternating subgroup. A function that is unchanged by the alternating subgroup is again of the form $A + B\Delta$ where A and B are symmetric and

$$\Delta = (x_1 - x_2)(x_1 - x_3) \ldots (x_{n-1} - x_n)$$

i.e., the product of *all* the differences $x_i - x_j$.

One can now try to find a function (like $x_1 + \omega x_2 + \omega^2 x_3$) whose power is unchanged by the alternating group. For $n = 4$ one can find a quadratic function of x_1,x_2,x_3,x_4 whose cube has this property. However for $n \geqslant 5$ it can be shown that *no such function exists,* and this is the first hint that there may be no way of solving equations of degree higher than four in terms of radicals. The proof of the nonexistence of a function whose power is unchanged by the alternating subgroup involves *only* the properties of the group of permutations of more than four objects.

To illustrate the versatility of the concepts of permutation and group we shall consider briefly the problem of classifying marriage laws in primitive societies. Through remarkable studies of several tribes (mainly by Lévi-Strauss and his collaborators) the following set of general rules was discovered:

(1) Every member of the tribe is assigned a *marriage type* and only individuals of the same type are allowed to marry.

(2) The type of an individual is determined *uniquely* by the individual's sex and by the type of the individual's parents.

(3) If two sets of parents are of different types, then their respective male offspring will be of different types. The same holds for female offspring.

(4) Whether a man is allowed to marry a female relative depends only on the manner in which they are related; in particular, no man is allowed to marry his sister.

(5) Some descendants of any two individuals must be allowed to marry.

It is clear that the marriage laws are determined by the knowledge of (a) how many marriage types $t_1, t_2, \ldots, t_n$ the given society uses and (b) by prescribing the rule that permits one to determine the type $S(t)(D(t))$ of the son (daughter) if t is the marriage type of the parents.

Rules 2 and 3 imply that $S(t)$ and $D(t)$ are permutations, and the first part of rule 4 implies that either $S(t)(D(t))$ is the same as t for every t (*i.e.*, the permutation is the *identity permutation*) or $S(t)(D(t))$ is different from t for every t (in which case it is called a *complete permutation*). A further consideration of the marriage rules leads to the conclusion that S and D must be so chosen that the group generated by them consist (except, of course, for the identity permutation I) of complete permutations only and that the group be *transitive*; *i.e.*, for any t_i and t_j there is a permutation P in the group such that $P(t_i) = t_j$. In Tarrau society, for example, $S(t) \equiv t$ while $D(t)$ is described as follows:

$$D = \begin{pmatrix} t_1 & t_2 & t_3 & t_4 \\ t_4 & t_1 & t_2 & t_3 \end{pmatrix}$$

Note that

$$D^2 = \begin{pmatrix} t_1 & t_2 & t_3 & t_4 \\ t_3 & t_4 & t_1 & t_2 \end{pmatrix}$$

$$D^3 = \begin{pmatrix} t_1 & t_2 & t_3 & t_4 \\ t_2 & t_3 & t_4 & t_1 \end{pmatrix}$$

$$D^4 = I$$

and that the group consists of I, D, D^2, and D^3. Another society also uses four marriage types but with

$$S = \begin{pmatrix} t_1 & t_2 & t_3 & t_4 \\ t_3 & t_4 & t_1 & t_2 \end{pmatrix} \qquad D = \begin{pmatrix} t_1 & t_2 & t_3 & t_4 \\ t_4 & t_3 & t_2 & t_1 \end{pmatrix}$$

Groups of permutations gave rise to a general and abstract theory of groups. An abstract group is a set of elements $f_0, f_1, f_2, \ldots$ in which we have a binary operation that associates with each pair of elements f_i and f_j a unique third element $f_i \cdot f_j$ which is also a member of the group. In addition the operation is assumed to have the following properties:

1. $(f_i \cdot f_j) \cdot f_k = f_i \cdot (f_j \cdot f_k)$ (associativity)
2. There is a unique *identity element* f_0 such that
$$f_0 \cdot f_i = f_i \cdot f_0 = f_i$$
3. Each element f_i has a unique inverse (denoted by f_i^{-1}); *i.e.*,
$$f_i \cdot f_i^{-1} = f_i^{-1} f_i = f_0$$

Even when defined abstractly a group can be thought of as a subgroup of the permutation group. In fact, if one thinks of the elements of the group as written out in some order $f_0, f_1, f_2, \ldots$, one can associate with each f_i the permutation

$$\begin{pmatrix} f_0, & f_1, & f_2, \ldots \\ f_i \cdot f_0, f_i \cdot f_1, f_i \cdot f_2, \ldots \end{pmatrix}$$

It may be checked that the permutations corresponding to $f_i \cdot f_j$ are the composition of the corresponding permutations (in proper order), and thus the original group and the corresponding group of permutations are indistinguishable (isomorphic).

The art of mathematical proof often consists in finding a framework in which what one is trying to prove becomes nearly obvious. Mathematical creativity consists largely in finding such frameworks. Sometimes one finds them in the rich world of material objects, sometimes (and this is the highest form of creativity) one invents them. More often than not, one recognizes that what one is interested in happens to fit into an already existing framework that was introduced originally for entirely different purposes. (When a framework is used repeatedly in different contexts, it becomes a theory and is studied for its own sake.)

A striking illustration is provided by applications of the group concept to elementary number theory. Consider, for example, the theorem of Wilson: if p is a prime number, then $(p - 1)! + 1$, *i.e.*, $1 \times 2 \times \ldots \times (p - 1) + 1$, is divisible by p. For example, if $p = 7$, $(p - 1)! + 1 = 721$ is divisible by 7.

There is nothing in the statement of the theorem to suggest that it has any connection with groups. However, with each prime number p one can associate in a very direct way a group of so-called *residues modulo p*. The elements of this group are the integers $1, 2, 3, \ldots, p - 1$, and the binary operation is defined as follows:

$$i \circ j = \text{remainder in dividing } i \times j \text{ by } p$$

For example, if $p = 7$ the "multiplication table" is

	1	2	3	4	5	6
1	1	2	3	4	5	6
2	2	4	6	1	3	5
3	3	6	2	5	1	4
4	4	1	5	2	6	3
5	5	3	1	6	4	2
6	6	5	4	3	2	1

and one can verify that the operation $\circ$ (called multiplication modulo p) satisfies all the needed conditions. In addition the group is also *commutative; i.e.,*

$$i \circ j = j \circ i$$

Consider now the expression

$$(1 \circ 2 \circ 3 \circ \cdots \circ (p-1))^2 = \underbrace{1 \circ 2 \circ 3 \circ \cdots \circ (p-1)}_{\text{I}} \circ \underbrace{1 \circ 2 \circ 3 \circ \cdots \circ (p-1)}_{\text{II}}$$

Because of commutativity we can pair off every element in I with its *inverse* in II and rewrite the "product" in that order. It then becomes clear that

$$(1 \circ 2 \circ 3 \circ \cdots \circ (p-1))^2 = 1$$

In other words, $1 \circ 2 \circ \cdots \circ (p-1)$ is an element whose square is the unit element.

If k is an element of the group such that

$$k \circ k = 1$$

it means that $k^2 - 1 = (k-1)(k+1)$ is divisible by p. Since $1 \leqslant k \leqslant p-1$, it follows that k is either 1 or $p-1$, and hence either

$$1 \circ 2 \circ 3 \circ \cdots \circ (p-1) = 1$$

or

$$1 \circ 2 \circ 3 \circ \cdots \circ (p-1) = p-1$$

If in the product $1 \circ 2 \circ 3 \circ \cdots \circ (p-1)$ we again pair off each element with its inverse, we will be left with $p-1$; *i.e.,* the only element except 1 that is its own inverse. Thus $1 \circ 2 \circ \cdots \circ (p-1) = p-1$, and this, in a slightly disguised form, is Wilson's Theorem.

Once one has thought of the group of residues modulo p, one is led to other number-theoretic facts. For example, there is a simple theorem that in a finite group (not necessarily commutative) of g elements, *every* element composed with itself g times is equal to the unit element:

$$\underbrace{f_i \circ f_i \circ \cdots \circ f_i}_{g} = f_0$$

If we apply this to the group of residues, for every k we obtain that $1 \leqslant k \leqslant p - 1$; or that

$$\underbrace{k \circ k \circ \cdots \circ k}_{p-1} = 1$$

or that $k^{p-1} - 1$ is divisible by p. This is a famous theorem announced by Fermat in 1640. Through the group concept it appears to be intimately related to the later theorem of Wilson.

In geometry and physics we again encounter the group concept. Transformations preserving distances and angles (rigid motion) form a group, and transformations of space-time leaving invariant "light-cones"

$$(x - x_0)^2 + (y - y_0)^2 + (z - z_0)^2 = c^2(t - t_0)^2$$

form the famous Lorentz group of special relativity.

Felix Klein in his celebrated Erlangen program proposed that we look upon geometry as a study of invariants of certain groups of transformations. Thus, according to Klein, Euclidean geometry is the study of invariants of the group that consists of translations, rotations, and reflections; projective geometry a study of invariants of the group of so-called projective transformations, and so on.

The fruitfulness of this point of view stems from the fact that algebraic properties of the group of transformations that leave a certain mathematical structure invariant reflect many of the properties of the structure itself.[15]

It is hard to exaggerate the role that the group concept plays in contemporary mathematics. There is not a corner in the whole subject that has not been significantly influenced by group-theoretic considerations in one way or another. In spite of the modesty of the axioms that define them, groups have yielded great mathematical riches, and more treasures remain hidden. Group theory is among the most vigorously studied parts of mathematics, and hardly a day passes without a new discovery or a new application.

[15] One even takes advantage of this point of view in the study of abstract groups by studying their so-called automorphisms. An *automorphism* of a group G is a transformation of G into itself that preserves the group operation; in other words, an automorphism f has the property that $f(p \cdot q) = f(p) \cdot f(q)$. The automorphisms of a given group G themselves form a group, the properties of which reflect many of the deeper structural properties of the original group G.

12a. HOMOLOGY GROUPS

An entirely different context in which the concept of a group plays a decisive role is provided by topology. Here we can give only the most elementary and cursory introduction to an important and widely studied theory known as *homology theory*.

The basic geometrical concepts of this theory are *simplices and simplicial complexes*. A tetrahedron is a three-dimensional simplex, a triangle is a two-dimensional simplex, an interval is a one-dimensional simplex, and a point is a zero-dimensional simplex. A complex is a collection of simplices such that any two are either disjoint or have a whole lower-dimensional simplex in common.

Fig. 10(a) is an example of a complex, while fig. 10(b) is a collection of simplices that do not form a complex since two pairs of triangles have only parts of their sides in common.

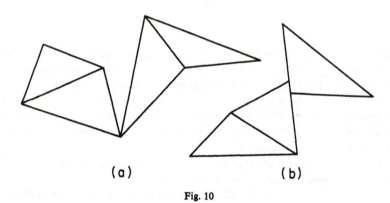

(a)   (b)

Fig. 10

A simplex is *oriented* if one assigns a *definite* ordering to its vertices. It thus may appear that there are (for example) six different ways of orienting a two-dimensional simplex corresponding to the six permutations of its three vertices; however, two orderings are considered equivalent if the first can be obtained from the second by an *even* permutation: a permutation decomposable into an *even* number of transpositions. For a triangle f_0, f_3, and f_4 are the even ones. Hence, there are only *two* different orientations. Zero-dimensional simplices obviously can be oriented in one way only. As will be seen below, it nevertheless will prove convenient to consider points themselves as having two orientations.

When a simplex has been oriented (*i.e.*, an ordering of its vertices has been decided upon), every one of its sub-simplices becomes automatically oriented too (induced orientation). The rule, however, is slightly different from what one might expect, and it reads as follows:

If a simplex has been oriented by the ordering $p_1, p_2, \ldots$ of its vertices, then the orientation of its face "opposite" p_k is that given by the ordering of the remaining p's if k is odd, and opposite to this natural orientation if k is even. A

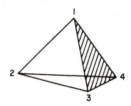

Fig. 11

similar definition is adapted for simplices of all dimensions. For example, if the tetrahedron in fig. 11 has been oriented (1 2 3 4), the shaded face does not become oriented (1 3 4) but (1 4 3) which we also write as $-(1\ 3\ 4)$.

This convention about induced orientations makes it necessary to assign two orientations to a point. Thus orientation (1 2) of a segment (one-dimensional simplex) induces orientation (1) but orientation (2 1) induces the opposite orientation $-(1)$ to the point 1.

When two oriented simplices have a simplex in common, they may induce on it orientations which are either the same or opposite.

Let $\sigma_1^{(k)}, \sigma_2^{(k)}, \ldots, \sigma_m^{(k)}$ be all the k-dimensional, oriented simplices of a given complex K. A k-dimensional chain is a formal expression of the form

$$a_1 \sigma_1^{(k)} + a_2 \sigma_2^{(k)} + \cdots + a_m \sigma_m^{(k)}$$

where $a_1, a_2, \ldots, a_m$ are integers (positive, negative, or zero). Chains of the same dimensionality may be "added" by adding coefficients of corresponding simplices, and with respect to this operation they form a group.

Given a k-dimensional, oriented simplex $\sigma^{(k)}$ we define as its boundary $\Delta(\sigma_i^{(k)})$ the chain

$$\sigma_{i1}^{(k-1)} + \sigma_{i2}^{(k-2)} + \cdots + \sigma_{ir}^{(k-r)} = \Delta(\sigma_i^{(k)})$$

where $\sigma_{i1}^{(k-1)}, \sigma_{i2}^{(k-2)}, \ldots, \sigma_{ir}^{(k-r)}$ are the $(k-1)$-dimensional simplices comprising the geometric boundary of σ_i, each taken with the orientation induced by the orientation of σ_i.

For example, the boundary of the tetrahedron in fig. 11 can be written as

$$- (1\ 2\ 3) - (1\ 3\ 4) + (1\ 2\ 4) + (2\ 3\ 4)$$

The boundary Δ of a chain $a_1\sigma_1^{(k)} + \ldots + a_m\sigma_m^{(k)}$ is defined by the formula

$$\Delta\,(a_1\sigma_1^{(k)} + \ldots + a_m\sigma_m^{(k)}) = a_1\,\Delta\,(\sigma_1^{(k)}) + \ldots + a_m\,\Delta\,(\sigma_m^{(k)})$$

except that if $\sigma_i^{(k)}$ and $\sigma_j^{(k)}$ have a $(k-1)$-dimensional simplex in common it will appear in the "sum" twice, once with the coefficient a_i and once with the coefficient a_j; the coefficient of this common simplex is then taken to be $a_i + a_j$ if the orientations agree, and $a_i - a_j$ if they disagree.

To calculate, for example, the boundary of the chain $-(1\ 2\ 3) - (1\ 3\ 4)$ $+(1\ 2\ 4) + (2\ 3\ 4)$, which as we have seen is itself the boundary of the simplex $(1\ 2\ 3\ 4)$, we note that

$$
\begin{aligned}
- \Delta\,(1\ 2\ 3) &= - (2\ 3) + (1\ 3) - (1\ 2) \\
- \Delta\,(1\ 3\ 4) &= - (3\ 4) + (1\ 4) - (1\ 3) \\
\Delta\,(1\ 2\ 4) &= (2\ 4) - (1\ 4) + (1\ 2) \\
\Delta\,(2\ 3\ 4) &= (3\ 4) - (2\ 4) + (2\ 3)
\end{aligned}
$$

and hence the required boundary is zero.

This is a special case of the general property that the boundary of a boundary is zero; symbolically

$$\Delta\Delta = 0$$

Consider now all chains B_r, the collection of all chains that are boundaries of $(r+1)$-dimensional chains and Z_r, the collection of all chains whose boundaries are zero (these are called cycles). Both are groups with respect to addition and the group of cycles Z_r contains the group B_r though it may be identical with it. The reason Z_r contains B_r is that $\Delta B_r = 0$ (the boundary of a boundary is zero) and hence each element of B_r is contained in the set of all r-dimensional chains that vanish (*i.e.*, that are equal to zero).

Let us now consider two cycles as equivalent if their difference is in B_r (in particular, all elements of B_r are considered equivalent). In this way the set Z_r breaks up into *disjoint* classes of equivalent cycles. This is a special case of a very general principle of "equivalence classes" discussed in greater detail in Chapter 2. A sum of two classes C_1 and C_2 is defined as follows: take a cycle c_1 from C_1 and a cycle c_2 from C_2; the sum $C_1 + C_2$ is defined as the class that contains $c_1 + c_2$. With respect to this new operation of addition, classes also form a group (the identity of this group is the whole class B_r) called the *factor group* of Z_r and B_r; it is denoted $H^{(r)}$ or Z_r/B_r;

$$H^{(r)} = Z_r/B_r$$

and is called the rth *homology group* of a complex.

Consider a two-dimensional complex consisting of the four faces of the tetrahedron in fig. 11 oriented (1 2 3), (1 2 4), (1 3 4), (2 3 4). Since our complex does not contain any three-dimensional simplices, B_2 consists only of the trivial element zero [16] and hence

$$H^{(2)} = Z_2$$

where Z_2 is the set of cycles, *i.e.*, the chains

$$a(1\ 2\ 3) + b(1\ 2\ 4) + c(1\ 3\ 4) + d(2\ 3\ 4)$$

for which

$$\Delta \{a(1\ 2\ 3) + b(1\ 2\ 4) + c(1\ 3\ 4) + d(2\ 3\ 4)\} = 0$$

We have as above

$$\Delta \{a(1\ 2\ 3) + b(1\ 2\ 4) + c(1\ 3\ 4) + d(2\ 3\ 4)\} =$$
$$a(2\ 3) - a(1\ 3) + a(1\ 2) + b(2\ 4) - b(1\ 4) + b(1\ 2)$$
$$+ c(3\ 4) - c(1\ 4) + c(1\ 3) + d(3\ 4) - d(2\ 4) + d(2\ 3) =$$
$$(a + d)\ (2\ 3) + (- a + c)\ (1\ 3) + (a + b)\ (1\ 2)$$
$$+ (b - d)\ (2\ 4) + (- b - c)\ (1\ 4) + (c + d)\ (3\ 4)$$

For this expression to vanish we must have

$$a + d = 0$$
$$- a + c = 0$$
$$a + b = 0$$
$$b - d = 0$$
$$- b - c = 0$$
$$c + d = 0$$

These equations are not independent since the last three can be derived from the first ones (*e.g.*, by adding the first two we obtain the sixth). From the first three we obtain

$$b = - a, \quad c = a, \quad d = - a$$

Hence the only two-dimensional cycles are chains of the form

$$a(1\ 2\ 3) - a(1\ 2\ 4) + a(1\ 3\ 4) - a(2\ 3\ 4)$$

With respect to the operation of addition the group of these chains is *indistinguishable* from the group of integers, and indeed may be identified with it. Thus

$$H^{(2)} = \text{group of integers with respect to addition}$$

By a slightly more involved computation we discover that $B_1 = Z_1 = $ group of integers with respect to addition, so that

$$H^{(1)} = Z_1/B_1 = \text{trivial group consisting only of the element } 0$$

---

[16] One might think that B_2 is empty; but it is more convenient to think of B_2 as the boundary of the chain 0 (1 2 3 4) that consists of the trivial element zero.

Suppose now that we take the complex consisting only of three faces (1 2 3), (1 2 4), (1 3 4) of the tetrahedron oriented as indicated. Now both $H^{(2)}$ and $H^{(1)}$ are trivial, each consisting only of the element zero.

On the other hand, if we took any closed convex polyhedron with triangular faces, its homology groups would be identical with those of the tetrahedron.

It appears that homology groups are related to the *intrinsic* way in which a complex is put together; this accounts for their importance and value in topology.

In topology two geometric configurations are considered identical if a one-to-one *continuous* correspondence between them can be established. Such a correspondence is called *homeomorphism,* and one can say that in topology configurations are identified that are homeomorphic. For example, topologically a tetrahedron and a sphere are identical. If a configuration can be approximated arbitrarily well by complexes, one can prove that all the approximating complexes have the same homology groups of all orders; therefore one can speak of homology groups of a configuration.

The fundamental theorem (going back to Poincaré) states that homeomorphic configurations have the same homology groups of all orders.

Let us illustrate the above discussion with a very simple example.

Consider fig. 12, a plane simple closed curve and fig. 13, a plane curve with one double point (a figure eight).

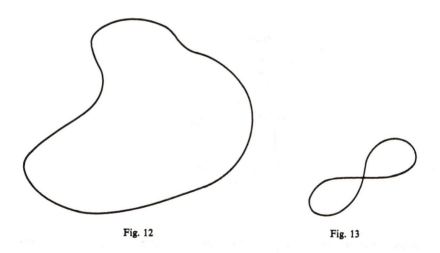

Fig. 12   Fig. 13

They are clearly not homeomorphic (for, on a simple plane curve the removal of any point leaves the rest connected; whereas the removal of a certain point—

the intersection—in the figure eight disconnects the remainder). Let us see how homology theory reflects this observation.

We can approximate the simple closed curve by *simple closed* polygons. A polygon is a one-dimensional complex (fig. 14),

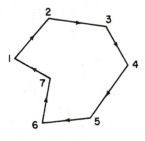

Fig. 14

and its homology group $H^{(1)}$ is the group of integers with respect to addition, *regardless of the number of sides*. Thus the one-dimensional homology group of a simple closed curve is the group of integers with respect to addition.

The curve of the figure eight also can be approximated this way but not by simple polygons (complexes).

A typical approximating complex is shown in fig. 15.

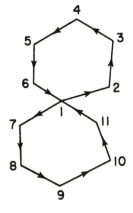

Fig. 15

The one-dimensional homology group is now the group of *ordered pairs* of *integers* (a,b) with respect to the operation of addition defined in the ordinary way; *i.e.,*

$$(a_1,b_1) + (a_2,b_2) = (a_1 + a_2, b_1 + b_2)$$

This group differs from the group of integers, so that a simple closed curve is not homeomorphic to the curve of figure eight.

While homeomorphic configurations have identical homology groups the converse does *not* hold for configurations of dimensionality higher than 2. The general problem of deciding if given higher-dimensional configurations are homeomorphic remains unsolved.

In this section we have seen an important example of a trend toward the *algebraization of mathematics*. It consists in approaching problems of geometry and analysis through the study of appropriately chosen algebraic structures (*e.g.*, groups) that are largely discrete and combinatorial in nature. This trend is characteristic of much of contemporary mathematics and we shall return to it in Chapter 2, where two more examples will be discussed.

13.   *Vectors, Matrices, and Geometry*

An important trend in mathematics has been the interpenetration and consequent unification of its apparently different parts. Analytic geometry, made possible by Descartes' introduction of coordinates into geometry, is a good example. By means of these coordinates, geometric objects such as conic sections can be expressed in algebraic equations and, by the same token, algebraic equations may be interpreted geometrically (for example, equations in two variables represent curves). The connection between algebra and geometry has been especially fruitful and far-reaching in modern times, and this is the theme of the present section. We shall see, for instance, how the familiar problem of solving simultaneous linear equations is interpreted geometrically. The abstract but elegantly simple concept of a linear vector space occupies a central position, and the principal developments involve what are called linear transformations of linear vector spaces. Linear transformations have an important concrete representation in terms of what are called matrices. These mathematical objects, all to be defined and discussed in this section, are simultaneously and inextricably algebraic and geometric, and so unite these disciplines.

As we have stressed, important ideas tend to find application in unexpected places; so it is with these ideas of linear algebra, as it has come to be called. In subsequent sections we shall see how they apply to the special theory of relativity and to Markov chains, an important topic in modern probability theory.

Throughout this study we have had several occasions to mention spaces of dimension higher than 3; perhaps it is worth stopping briefly to discuss these objects and some closely related topics in some detail.

Let us first introduce the concept of a *linear vector space*.

To begin with consider the familiar plane. Pick a point 0 in it (fig. 16) and fix it once and for all. With each point *P* in the plane we associate a *directed* line segment (vector) from 0 to *P*.

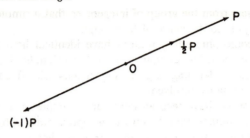

Fig. 16

In this way points become identified with vectors. We now introduce two important operations on vectors.

1. Multiplication by a real number (scalar) α is defined as follows: αP is the vector that is $|\alpha|$ times longer than P along the line of P. (Recall that $|\alpha|$ denotes the absolute value of α; e.g., $|-2| = 2$, $|2| = 2$). This vector points the same way as P if $\alpha > 0$ and the opposite way if $\alpha < 0$.

2. Addition of vectors is defined by the familiar parallelogram construction illustrated in fig. 17.

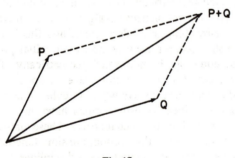

Fig. 17

It now may be verified that the two basic operations have the following properties:

(*a*). $P + Q = Q + P$
(*b*). $(P + Q) + R = P + (Q + R)$
(*c*). There exists a unique vector 0 such that for every vector P, $P + 0 = P$.
(*d*). For every vector P there exists a unique vector P' such that $P' + P = 0$.
(*e*). $\alpha(P + Q) = \alpha P + \alpha Q$
(*f*). $(\alpha + \beta)P = \alpha P + \beta P$
(*g*). $\alpha(\beta P) = (\alpha\beta)P$
(*h*). $(1)P = P$

We can now reverse our procedure (as one often does in mathematics) and define a linear vector space to be a set of objects P, Q, . . . on which operations of scalar multiplication and addition are defined and which satisfy the above eight properties (axioms).

Although these axioms were suggested by consideration of plane vectors, they are also satisfied (*e.g.*) by three-dimensional vectors and by objects that have little in common with vectors. For example, polynomials with real coefficients form a linear vector space; real-valued continuous functions defined on an interval as well as chains defined in Section 12a are also examples of vector spaces.

The fact that such different objects as plane (or space) vectors and continuous functions are examples of linear vector spaces shows how little specificity of structure is implied by axioms (*a*) through (*h*).

It will be seen that one can enrich the structure by imposing additional axioms.

Let us now show how one can introduce the concept of dimensionality. One begins by introducing the notion of *linear independence*.

Vectors $P_1, P_2, \ldots, P_n$ are called linearly independent if a linear relation between them of the form

$$\alpha_1 P_1 + \alpha_2 P_2 + \cdots + \alpha_n P_n = 0$$

implies that $\alpha_1 = \alpha_2 = \ldots = \alpha_n = 0$. In other words, no P_i can be written as a linear expression (linear combination) involving only the remaining P's.

A linear vector space is said to be of dimension n if there exist n linearly independent vectors $P_1, P_2, \ldots, P_n$, but *there is no collection of* $(n + 1)$ *linearly independent vectors.*

If a vector space is n-dimensional, the n linearly independent vectors $P_1, P_2, \ldots, P_n$ are said to form a *basis* (or base). This means that vector P can be written in a unique way as a linear combination of $P_1, P_2, \ldots, P_n$

$$P = \alpha_1 P_1 + \cdots + \alpha_n P_n$$

Having chosen a basis, we see that to every vector P there corresponds in a unique way an ordered n-tuple of scalars $(\alpha_1, \alpha_2, \ldots, \alpha_n)$ and *vice versa*. We also call the basis $(P_1, \ldots, P_n)$ a coordinate system and the α's the coordinates of P.

A linear transformation T of a vector space into itself is a mapping that associates with each vector P another vector $T(P)$ in the same space and that has the following properties:

(*a*) $$T(P + Q) = T(P) + T(Q)$$
(*b*) $$T(\alpha P) = \alpha T(P)$$

Using a basis (coordinate system) we can associate with each linear transformation in a unique way a square array of scalars known as a *matrix*.

To see how this can be done let us choose a basis $P_1, P_2, \ldots P_n$ and consider the "transformed" vectors $T(P_1), T(P_2), \ldots, T(P_n)$. Since each of these transformed vectors can be expressed as a linear combination of the original vectors of the basis, we can write

$$T(P_1) = t_{11}P_1 + t_{12}P_2 + \cdots + t_{1n}P_n$$
$$T(P_2) = t_{21}P_1 + t_{22}P_2 + \cdots + t_{2n}P_n$$
$$\cdots\cdots\cdots\cdots\cdots\cdots\cdots\cdots\cdots\cdots\cdots$$
$$T(P_n) = t_{n1}P_1 + t_{n2}P_2 + \cdots + t_{nn}P_n$$

Thus with the linear transformation T there is associated (in a unique way once a basis has been selected) the matrix of scalars:

$$T = \begin{pmatrix} t_{11} & t_{12} & \ldots & t_{1n} \\ t_{21} & t_{22} & \ldots & t_{2n} \\ \cdots & \cdots & \cdots & \cdots \\ t_{n1} & t_{n2} & \ldots & t_{nn} \end{pmatrix}$$

Suppose we have a second linear transformation S whose matrix with reference to the basis $P_1, P_2, \ldots, P_n$ is

$$S = \begin{pmatrix} s_{11} & s_{12} & \ldots & s_{1n} \\ s_{21} & s_{22} & \ldots & s_{2n} \\ \cdots & \cdots & \cdots & \cdots \\ s_{n1} & s_{n2} & \ldots & s_{nn} \end{pmatrix}$$

and imagine that we first apply the transformation T and then the transformation S. The combined action of the two transformations is the composition ST which again is easily seen to be a linear transformation. The matrix associated with ST (again with respect to the same basis $P_1, P_2, \ldots, P_n$) is the matrix whose (i,k) entry (ith row, kth column) is given by the formula

$$s_{i1}t_{1k} + s_{i2}t_{2k} + \cdots + s_{in}t_{nk} = \sum_{j=1}^{n} s_{ij}t_{jk}$$

If we composed S and T in the opposite order (*i.e.*, first S then T) then the (i,k) elements of the matrix corresponding to TS are

$$t_{i1}s_{1k} + \cdots + t_{in}s_{nk} = \sum_{j=1}^{n} t_{ij}s_{jk}$$

which, in general, is different from

$$\sum_{j=1}^{n} s_{ij}t_{jk}$$

In this way *composition* of linear transformations is reflected in an operation, called *multiplication*, of corresponding matrices.

The identity transformation *I* is represented (regardless of base) by the so-called unit or identity matrix.

$$I = \begin{pmatrix} 1 & 0 & 0 & \ldots & 0 \\ 0 & 1 & 0 & \ldots & 0 \\ 0 & 0 & 1 & \ldots & 0 \\ \ldots & \ldots & \ldots & \ldots \\ 0 & 0 & \ldots & 0 & 1 \end{pmatrix}$$

One can also attempt to define an inverse of a transformation *T* as a transformation that, when composed with *T*, yields the identity transformation *I*. However, not every linear transformation *T* has an inverse. Those that do not are characterized by the property that they annihilate a nonzero vector. (The image of the vector under *T* is the zero vector.) In other words a linear transformation *T* has an inverse T^{-1} (which then also is unique) if and only if $T(P) = 0$ implies that *P* is the zero vector. The composition of two nonsingular transformations is again such—they form a group.

Let us discuss this point in a little more detail in the case of a two-dimensional real linear vector space.

If we choose a base once and for all, vectors become identified with ordered pairs (α_1, α_2) of real numbers and linear transformations with 2×2 matrices of real numbers:

$$\begin{pmatrix} t_{11} & t_{12} \\ t_{21} & t_{22} \end{pmatrix}$$

The statement that *T* has an inverse is tantamount to saying that the only solution of the system of linear equations

$$t_{11}\alpha_1 + t_{12}\alpha_2 = 0$$
$$t_{21}\alpha_1 + t_{22}\alpha_2 = 0$$

is the trivial result $\alpha_1 = \alpha_2 = 0$.

Now we are on the familiar ground of high-school algebra, and we can show that this will be the case if and only if

$$t_{11}t_{22} - t_{21}t_{12} \neq 0$$

The quantity $t_{11} t_{22} - t_{21} t_{12}$ is called the *determinant* of the matrix *T*, and in high school one used the notation

$$t_{11}t_{22} - t_{12}t_{21} = \det T = \begin{vmatrix} t_{11} & t_{12} \\ t_{21} & t_{22} \end{vmatrix}$$

It now can be shown that the inverse matrix T^{-1} is given by the formula

$$T^{-1} = \begin{pmatrix} \dfrac{t_{22}}{\det T} & -\dfrac{t_{12}}{\det T} \\ -\dfrac{t_{21}}{\det T} & \dfrac{t_{11}}{\det T} \end{pmatrix}$$

i.e.,

$$T^{-1}T = TT^{-1} = I = \begin{pmatrix} 1 & 0 \\ 0 & 1 \end{pmatrix}$$

Can one extend the concept of the determinant to n dimensions, and can one generalize the above formula for the inverse of a 2×2 matrix to the $n \times n$ case?

The original motivation for the consideration of such questions was provided by a desire to solve systems of n linear equations with n unknowns $\alpha_1, \alpha_2, \ldots \alpha_n$.[17]

$$
\begin{aligned}
t_{11}\alpha_1 + t_{12}\alpha_2 + \cdots + t_{1n}\alpha_n &= \beta_1 \\
t_{21}\alpha_1 + t_{22}\alpha_2 + \cdots + t_{2n}\alpha_n &= \beta_2 \\
\text{------------------------} \\
t_{n1}\alpha_1 + t_{n2}\alpha_2 + \cdots + t_{nn}\alpha_n &= \beta_n
\end{aligned}
$$

For $n = 2$, 3, and even 4 one can write down $\alpha_1, \alpha_2, \ldots$ as complicated-looking ratios of certain algebraic forms involving the t's and β's. It is possible to guess and then to prove that

$$\alpha_k = \frac{\det T_k}{\det T}$$

where the matrix T_k is obtained by replacing the kth column of T by $\beta_1, \beta_2, \ldots, \beta_n$, and where the determinant of general $n \times n$ matrix

$$A = \begin{pmatrix} a_{11} & a_{12} & \ldots & a_{1n} \\ a_{21} & a_{22} & \ldots & a_{2n} \\ \cdots\cdots\cdots\cdots \\ a_{n1} & a_{n2} & \ldots & a_{nn} \end{pmatrix}$$

is calculated by the following rule:

Let

$$\pi = \begin{pmatrix} 1 & 2 & 3 & \ldots & n \\ i_1 & i_2 & i_3 & \ldots & i_n \end{pmatrix}$$

---

[17] We hope that the reader will not be unduly perturbed by our departure from the convention of using x, y, and z to denote "unknown" quantities. While such conventions are often convenient they are seldom binding and it is well to be reminded once in a while that they are only conventions.

be a permutation of the indices $1, 2, \ldots, n$, and associate with this permutation the product

$$\pm\, a_{1i_1} a_{2i_2} \ldots a_{ni_n}$$

with the sign $+$ if the permutation π is even and the sign $-$ if the permutation π is odd.

(As mentioned before, every permutation is a product of *transpositions; i.e.*, permutations that exchange two elements and leave all others unaltered. While such a decomposition is not unique the number of transpositions involved is either even or odd. In the first case the permutation is called even and in the other odd.)

The determinant is the sum of these signed products taken over all $(n!)$ permutations:

$$\det A = \sum_{\pi} \operatorname{sgn} \pi\; a_{1\pi(1)}\, a_{2\pi(2)} \ldots a_{n\pi(n)}$$

We have used sgnπ for $+1$ if π is even and -1 if π is odd and $\pi(1), \pi(2), \ldots$ for $i_1, i_2, \ldots$

Introduced in this way the determinant appears as a complicated algebraic construct whose main usefulness is in solving systems of linear equations. (The actual calculation of determinants seldom directly uses the definition. There is a vast literature on properties of determinants and on methods of computing them.) But in the context of linear vector spaces we can view determinants from a more geometric point of view.

To see how this comes about consider a basis $P_1, P_2, \ldots, P_n$ and a linear transformation T of our vector space into itself.

Consider now another basis $Q_1, Q_2, \ldots, Q_n$. It is reasonable to inquire how the representations of vectors and matrices *transform* from one basis to another.

Since the vectors P_j form a basis, we can represent each Q_i as a linear combination of the P's; *i.e.*,

$$Q_i = d_{i1}P_1 + d_{i2}P_2 + \ldots + d_{in}P_n = \sum_{j=1}^{n} d_{ij}P_j$$

and we see that going from one basis to another can be described by the matrix

$$D = \begin{pmatrix} d_{11} & d_{12} & \ldots & d_{1n} \\ d_{21} & d_{22} & \ldots & d_{2n} \\ \multicolumn{4}{c}{\dotfill} \\ d_{n1} & d_{n2} & \ldots & d_{nn} \end{pmatrix}$$

On the other hand, since we have assumed that the Q's also form a basis, we have

$$P_i = \sum_{j=1}^{n} c_{ij} Q_j$$

which defines the matrix C

$$C = \begin{pmatrix} c_{11} & c_{12} & \dots & c_{1n} \\ c_{21} & c_{22} & \dots & c_{2n} \\ \dots\dots\dots\dots\dots \\ c_{n1} & c_{n2} & \dots & c_{nn} \end{pmatrix}$$

It now can be checked that

$$CD = DC = I$$

so that $D = C^{-1}$ and $C = D^{-1}$ or, in words, the matrices are inverses of each other. We can summarize all this by saying that n linear combinations of vectors forming a basis, themselves form a basis if and only if the matrix D of coefficients is nonsingular; *i.e.*, if D^{-1} exists, or equivalently, det $D \neq 0$.

If now

$$T = \begin{pmatrix} t_{11} & t_{12} & \dots & t_{1n} \\ t_{21} & t_{22} & \dots & t_{2n} \\ \dots\dots\dots\dots\dots \\ t_{n1} & t_{n2} & \dots & t_{nn} \end{pmatrix}$$

is the matrix associated with the linear transformation T with respect to the basis $(P_1, \dots, P_n)$ it can be shown that the matrix

$$C^{-1} TC = DTD^{-1}$$

is the matrix associated with the *same* transformation but with respect to the basis $(Q_1, \dots, Q_n)$.

Algebraically the matrices $C^{-1} TC$ (as C runs through all $n \times n$ matrices) that have an inverse (such matrices are called nonsingular) will be quite different in appearance. But since we *know* that *geometrically* they all describe the *same* linear transformation T, they all must have something in common. In particular, one may inquire whether, using entries (elements) of a matrix, one can construct expressions that remain unchanged (invariant) when the matrix is postmultiplied by C and premultiplied by its inverse C^{-1}. It turns out that the determinant is such an expression:

$$\det (C^{-1} TC) = \det (T)$$

and hence it is connected with the transformation in an *intrinsic* manner.

It therefore must be possible to define the determinant associated with the linear transformation T without reference to a coordinate system (basis). This is indeed the case, and it can be done as follows.

One begins with the search for a (nonvanishing) vector E that transforms into a scalar multiple of itself under the transformation T.

In other words one seeks vectorial solutions of the equation $TE = \lambda E, E \neq 0$.

One can show that in an n-dimensional vector space there are n linearly independent vectors $E_1, E_2, \ldots, E_n$ of this kind. They are called *eigenvectors* of the transformation T and the corresponding scalars $\lambda_1, \lambda_2, \ldots, \lambda_n$ are called the *eigenvalues*.

It is clear that in the coordinate system $E_1, E_2, \ldots, E_n$ the matrix description of T is particularly simple; it is diagonal:

$$\begin{pmatrix} \lambda_1 & 0 & 0 & \ldots & 0 \\ 0 & \lambda_2 & 0 & \ldots & 0 \\ 0 & 0 & \lambda_3 & \ldots & 0 \\ \multicolumn{5}{c}{\ldots\ldots\ldots\ldots\ldots} \\ 0 & 0 & 0 & \ldots & \lambda_n \end{pmatrix}$$

The determinant of a diagonal matrix is the product of the diagonal elements and we have

$$\det T = \lambda_1 \cdot \lambda_2 \ldots \lambda_n$$

which may be taken as an algebraic definition of the determinant. This *number*, associated with a matrix, indicates some of the most important properties of the transformations it describes. For example, a linear transformation T has an inverse if and only if $\det T \neq 0$.

We shall return to the determinant a little later in this section but first we must put more structure into our linear vector space.

So far we have not mentioned the words *distance* and *angle*, except in defining αP. (Even there it was not absolutely necessary. We could have defined nP, for a positive integer n, as $P + P + \cdots + P$ taken n times and $\frac{1}{n} P$ as that vector Q which added to itself n times gives P. In this way we could also define $\frac{m}{n} P$ where both m and n are positive integers. Having thus defined rP for positive rational r we could define $-rP$ as that unique vector which when added to rP gives 0. To extend all this to real multiples of P we could follow the procedure, described briefly in Section 7, of extending rationals to reals.)

Let us recall how these notions are handled in elementary plane analytic geometry.

There we begin with two perpendicular lines

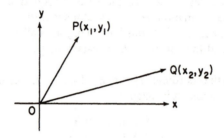

Fig. 18

(the x-axis and the y-axis) and assign to each point P coordinates x and y as shown on fig. 18.

Using the Pythagorean Theorem it is then proved that the distance between points $P(x_1, y_1)$ and $Q(x_2, y_2)$ is given by the formula

$$d(P, Q) = \sqrt{(x_2 - x_1)^2 + (y_2 - y_1)^2}$$

Using the Law of Cosines (a slight extension of the Pythagorean Theorem) it is also shown that the angle θ between OP and OQ can be determined from the formula

$$\cos \theta = \frac{x_1 x_2 + y_1 y_2}{\sqrt{x_1^2 + y_1^2} \sqrt{x_2^2 + y_2^2}}$$

Somewhat more generally the cosine of the angle between the segment AP and AQ where (fig. 19) the coordinates of A are x_0 and y_0 is given by similar formula

$$\cos \theta = \frac{(x_1 - x_0)(x_2 - x_0) + (y_1 - y_0)(y_2 - y_0)}{\sqrt{(x_1 - x_0)^2 + (y_1 - y_0)^2} \sqrt{(x_2 - x_0)^2 + (y_2 - y_0)^2}}$$

One now can try to reverse the usual procedure and take the above formulas as definitions of distance and angle.

But now there arise two problems:

1. Since the formulas refer to a particular coordinate system, one must check that they do not change in appearance by going over to another coordinate system of the same kind.

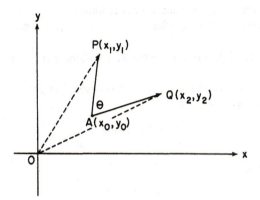

Fig. 19

In other words, if we choose a different system (fig. 20) of mutually perpendicular axes (*x*-axis and *y*-axis), and if in this new system the coordinates of *P* are (x_1', y_1'), of $Q(x_2', y_2')$, and of 0 (x_0', y_0'), then one must have

$$(x_2' - x_1') + (y_2' - y_1')^2 = (x_2 - x_1)^2 + (y_2 - y_1)^2$$

and a similar identity coming from the formula for cos θ.

2. How does one define *congruence* of geometric configurations, a concept that is central to the whole development of geometry?

The two problems are closely related, and the answer to the second will almost automatically provide us with an answer to the first.

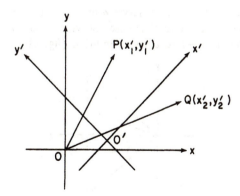

Fig. 20

Behind the notion of congruence lies the intuition concerning *rigid motions;* *i.e.*, motions that do not change the *metric* relationships of geometric configurations.

More precisely, rigid motions are transformations that preserve distances and angles.

It is possible to show that every transformation that preserves distances and cosines of angles is of the form

$$x' = a_{11}x + a_{12}y + x_0$$
$$y' = a_{21}x + a_{22}y + y_0$$

where the matrix

$$A = \begin{pmatrix} a_{11} & a_{12} \\ a_{21} & a_{22} \end{pmatrix}$$

is such that $A'A = I$ and

$$A' = \begin{pmatrix} a_{11} & a_{21} \\ a_{12} & a_{22} \end{pmatrix}$$

is the transpose of A; *i.e.*, the matrix obtained by reflecting A across its main diagonal (upper left to lower right).

It can be checked that det A is either $+1$ or -1. If det $A = -1$, the transformation (though conserving distances and *cosines* of angles) does not represent a rigid motion in the plane.

For example, the transformation

$$x' = x$$
$$y' = -y$$

is a reflection in the *x*-axis that conserves d and cos θ but reverses the orientation (from counterclockwise to clockwise and vice versa). In general a transformation with det $A = -1$ reverses orientation (fig. 21).

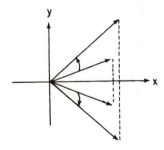

Fig. 21

Rigid motions are thus transformations of the form

$$x' = a_{11}x + a_{12}y + x_0$$
$$y' = a_{21}x + a_{22}y + y_0$$

with $A'A = I$ and det $A = 1$.

It turns out that changing to a different (rectilinear) coordinate system is described by formulas of the same general form as those describing rigid motions, and problem 1 is thus solved.

Now that we have an *explicit* and *concrete* definition of congruence, we must connect it with Euclid's concept of congruence.

Euclid does not define congruence and does not tell us what it is. Instead, he merely lists (in the form of axioms) the properties it *must have* to satisfy intuition.

On the basis of these and other axioms (including the famous Fifth Postulate that through a point outside a line *l* there is one and only one line parallel to *l*) one could prove the Pythagorean Theorem in the form that involves only the concept of *congruence*. Such a proof is illustrated in fig. 22 where the square based on the hypotenuse of a right triangle and the figure composed of the two squares based on the legs with the smaller on top of the bigger one can each be decomposed into five mutually congruent pieces (four congruent copies of the original triangle and a square).

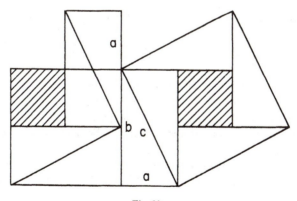

Fig. 22

To get from this *nonnumerical* version of the Pythagorean Theorem the familiar *numerical* version $c^2 = a^2 + b^2$ one needs a *theory of measuring* that again can be based on the axioms, none of which involves numbers as such. Here the Greeks encountered difficulty with irrational numbers because the axioms they used allowed them to construct only rational numbers.

The fact (following from the Pythagorean Theorem) that the hypotenuse of a right isosceles triangle is incommensurable with its leg (irrationality of $\sqrt{2}$) was a source of deep concern to the Greeks. We know now that to complete the theory of measuring to allow irrational as well as rational numbers, one needs a new and subtle axiom, the so-called *axiom of continuity*. Euclid and the Greeks also missed a whole group of axioms called axioms of order. These axioms concern the notion of a point lying between two other points, a half-line OA lying between two other half-lines OP and OQ, and so on. They express properties intuitively so obvious it is no wonder they were overlooked. Still, a computing machine, not being able to "see" whether a point is or is not between two other points, must be "told" how to handle the concept of "betweenness."

Again one need not *know* what "to lie between" means as long as one has a sufficiently complete list of properties of "lying between."

One must therefore check that our algebraically defined congruence has all the properties postulated by Euclid. Once this is done and once all other axioms have been checked, we have a complete algebraic (or analytic) *model* of plane Euclidean geometry.

It is now easy to go beyond two dimensions, and this is best done in the framework of linear vector spaces.

The only new concept one needs is that of a *scalar product* $\omega(P,Q)$ of two vectors P and Q.

This is introduced by postulating the existence of $\omega(P,Q)$ with the following properties:

(a) $\omega(P, Q)$ is a symmetric bilinear function of P and Q; i.e., $\omega(P, Q) = \omega(Q, P)$ and
$$\omega(P, \alpha Q + \beta R) = \alpha\omega(P, Q) + \beta\omega(P, R)$$
(b) $\omega(P, P) \geqslant 0$ and $\omega(P, P) = 0$ only if $P = 0$.

Let us show that a two-dimensional vector space on which a scalar product $\omega(P,Q)$ with the above properties is defined can be made into a model of the Euclidean plane.

Take two linearly independent vectors P_1 and P_2.

It can be shown that linear combinations E_1 and E_2 of P_1 and P_2 can be found such that
$$\omega(E_1, E_1) = \omega(E_2, E_2) = 1$$
and
$$\omega(E_1, E_2) = 0$$
In the proof one needs the fact that if P_1 and P_2 are linearly independent then

$$\omega(P_1, P_1)\omega(P_2, P_2) - \omega^2(P_1, P_2) > 0$$

This is a form of the famous inequality of Schwarz.

It follows that E_1, E_2 also form a basis and writing

$$P = x_1E_1 + y_1E_1$$
$$Q = x_2E_1 + y_2E_2$$

we obtain

$$\omega(P, Q) = x_1x_2 + y_1y_2$$

If one now defines the square of the distance between vectors P and Q by the formula

$$d^2(P, Q) = \omega(P - Q, P - Q)$$

and the cosine of the angle between P and Q by the formula

$$\cos \theta = \frac{\omega(P, Q)}{\sqrt{\omega(P, P)} \sqrt{\omega(Q, Q)}}$$

one is back to the familiar formulas of plane analytic geometry.

We can now define the n-dimensional Euclidean space as *an n-dimensional real vector space with a scalar product* $\omega(P,Q)$ *that has the properties* (*a*) *and* (*b*) *stated above.*

Rigid motions are *translations* and *linear transformations* T (called by analogy *rotations*) that preserve the scalar product,

$$\omega(TP, TQ) = \omega(P, Q)$$

and that also are orientation-preserving; *i.e.*, det T = 1. The conclusion that det $T = \pm1$ follows from $\omega(TP,TQ) = \omega(P,Q)$.

Rotations and translations form a group, and Euclidean geometry (in any number of dimensions) can be defined (following Felix Klein's Erlangen Program) as the study of properties of configurations left invariant by this group.

Translations and all nonsingular linear transformations (*i.e.*, those that possess unique inverses) also form a group much larger than the Euclidean group. The properties left invariant by this larger group are fewer, and they form the subject of so-called *affine geometry* (the corresponding group is called the affine group).

In this geometry there is no way to distinguish between a circle and an ellipse (or between two ellipses) because a circle can be transformed into an ellipse by affine transformation, and only properties that do not change under such transformations are the legitimate concern of affine geometry. However, it is possible to distinguish between a hyperbola and an ellipse since one cannot be transformed into another by an affine transformation.

Projective geometry is even more primitive. Since it studies properties left invariant by projections, it cannot even distinguish between ellipses and hyperbolas because, being conic sections, either is obtainable from the other by a projection.

Returning to n-dimensional Euclidean geometry, let us conclude with a few remarks.

Let $E_1, E_2, \ldots, E_n$ be an orthonormal basis; *i.e.*,

$$\omega(E_i, E_i) = 1 \qquad i = 1, 2, \ldots, n$$
$$\omega(E_i, E_j) = 0 \qquad i \neq j$$

In other words an orthonormal basis is a set of n *mutually orthogonal* (perpendicular) *unit* vectors. The existence of such a basis must and can be demonstrated. The elements of such a basis are linearly independent.

A unit cube based on $E_1, E_2, \ldots, E_n$ is the set of vectors

$$P = x_1 E_1 + \cdots + x_n E_n$$

such that

$$0 < x_i < 1 \qquad i = 1, 2, \ldots, n$$

More generally, a "rectangular parallelepiped" based on $E_1, E_2, \ldots, E_n$ is the set of vectors $P = x_1 E_1 + \cdots + x_n E_n$ such that

$$0 < x_1 < a_1, \qquad 0 < x_2 < a_2, \qquad \ldots \qquad 0 < x_n < a_n$$

The volume of such a parallelepiped is defined to be $a_1 a_2 \ldots a_n$ (so that in particular the volume of the unit cube is equal to unity).

Having assigned volumes to "rectangular parallelepipeds" by using the axiom of additivity (and complementarity) we can extend the concept of volume (or better, n-dimensional Lebesgue measure) to a vast collection of sets. In fact, it can be shown that our choice of $a_1 a_2 \ldots a_n$ as the volume of the rectangular parallelepiped is the only one consistent with (a) the axiom of additivity, (b) the requirement that congruent sets be assigned the same measure, and (c) the requirement that the volume vary continuously as the lengths of the sides do.

If we subject our unit cube to a linear transformation T, we obtain a parallelepiped that, in general, will be "skew" (*i.e.*, not rectangular). Its volume turns out to be $\pm \det T$, where the sign plus or minus is chosen to make the complete expression positive.

In general if a set Ω is subject to a linear transformation T, the measure of the transformed set $T(\Omega)$ is obtained by multiplying the measure of Ω by the determinant of T with an appropriate sign.

Symbolically,

$$m(T(\Omega)) = \pm \det T m(\Omega)$$

The strange algebraic form

$$\sum_\pi \operatorname{sgn} \pi \; t_{1\pi(1)} \, t_{2\pi(2)}, \; \ldots \, t_{n\pi(n)}$$

to which one is led by solving systems of linear equations emerges in a new highly appealing, geometric light.

Many properties of determinants that are obtained with great labour by algebraic manipulations with matrices become nearly self-evident once the geometric interpretation becomes available.

As an example let us mention the theorem (used implicitly several times throughout this section) that the determinant of a product of two matrices is the product of determinants.

$$\det (TS) = \det T \det S$$

An algebraic proof is laborious and obscures the meaning of the theorem. Geometrically, the theorem is obvious, since it merely states that the distortion of volume produced by applying in succession two linear transformations is the product of individual distortions.

One should not, however, be led to believe that one can get "something for nothing."

The theorem about the determinant of the product becomes obvious only *after* we have proved the distortion theorem

$$m(T(\Omega)) = \pm \det Tm(\Omega)$$

and this theorem is neither obvious nor immediate!

A case can be made against the geometric proof on the grounds that too much extraneous material has to be introduced before the proof can be made intelligible.

One might well ask, why bother about n-dimensional parallelepipeds and their volumes if one can give a proof by elementary (if somewhat tedious) algebraic manipulations?

In a way, such questions cannot be answered. Logically, a proof is a proof, and the validity of a theorem is independent of the way it is proved (provided only that one adheres strictly to the logical rules of the game).

Fortunately, as we have stressed repeatedly, there is more to mathematics than mere logic. The "flavour" of a theorem depends largely on the context in which it is formulated even though its truth does not. It is the context that makes it possible to distinguish between puzzles and problems and between an assembly of accidental facts and a coherent theory.

14. *Special Theory of Relativity as an Example of the Geometric View in Physics*

Let us now show how some of the concepts discussed in the preceding section can be modified and extended to provide a mathematical framework for the special theory of relativity.

The special theory of relativity was conceived by Albert Einstein as a reconciliation of Newtonian mechanics with the wave theory of light.

Briefly, the dilemma was as follows: In Newtonian mechanics all observers moving uniformly and rectilinearly with respect to each other are equivalent. If each such observer kept a notebook in which he recorded the results of his observations and measurements of *mechanical phenomena* and if the notebooks were later compared, they would be identical for all intents and purposes.

On the other hand, the wave theory of light required a medium in which light could propagate and such a medium, called luminiferous ether, was in fact postulated. Ether provided a preferred frame of reference, and it appeared that it should be possible to detect uniform motion with respect to ether by means of light signals. In particular, it should be possible to detect the motion of the earth through ether by comparing the time it takes a light signal first to travel a distance l and then back in the direction of motion and then in the direction perpendicular to the motion. Though the time difference would be of the tiny order $(v/c)^2$ (where v is the velocity of the earth and c is the velocity of light), it could be detected by means of a very accurate interferometer.

The experiment to detect this time difference was performed in 1887 by Michelson and Morley, and the result was negative! The negative outcome of the Michelson-Morley experiment precipitated a crisis in physics that seemed to be fully resolved only in 1905 by Einstein's special theory of relativity.

Einstein proposed to keep the equivalence of observers moving uniformly and rectilinearly with respect to each other (the principle of relativity). He also proposed that the result of the Michelson-Morley experiment become a new law, to the effect that the velocity c of propagation of light in vacuum be the same for all these observers.

To combine into a harmonious whole the principle of relativity with the principle of constancy of the velocity of light propagation required a deep revision of our concepts of space and time. To understand this revision and to catch a glimpse of some of its implications, it is best to follow a geometric procedure that we owe to Hermann Minkowski.

Imagine a rectangular coordinate system S and another such system S' which moves with constant (uniform) velocity v in the direction of the positive x-axis (fig. 23).

Let O be an observer at rest with respect to S and O' an observer at rest with respect to S'. Each of these observers has his own measuring rods, and in each system we can imagine synchronized clocks placed as densely throughout the system as needed.

An event to observer O is an (ordered) quadruplet (x, y, z, t) of real numbers, the first three indicating *where,* and the last one *when,* the event in question occurred. The same event to observer O' will be a different set of four numbers (x', y', z', t') that he will obtain by using *his* rods and *his* clocks.

Now the question is, how are the two quadruplets related?

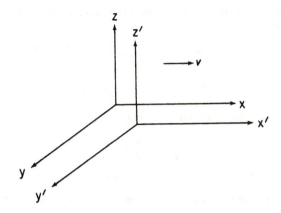

Fig. 23

If one takes the Newtonian point of view that time is absolute, then the clocks in the two systems could be synchronized and we would have

$$t' = t$$

In addition, we would clearly have

$$y' = y$$
$$z' = z$$
$$x' = x - vt$$

This transformation from unprimed coordinates (x, y, z, t) to primed ones (x', y', z', t') is called Galilean, and all the laws of classical dynamics remain invariant under it.

However, if a flash of light is produced by O at $(0, 0, 0, 0)$ (when we assumed that both systems coincide) the wavefront moves through space according to the law

$$x^2 + y^2 + z^2 = c^2t^2$$

The observer O' who watches the light signal propagate must, according to Einstein, arrive at the same equation in primed coordinates; *i.e.,*

$$x'^2 + y'^2 + z'^2 = c^2t'^2$$

and this is *incompatible with the Galilean transformation.*

Let us seek the transformation from primed to unprimed variables that yields

$$x'^2 + y'^2 + z'^2 = c^2t'^2$$

whenever

$$x^2 + y^2 + z^2 = c^2t^2$$

First, by analogy with the Galilean transformation, we assume that the new transformation is also linear.[18] Next one can convince oneself that one actually has the identity

$$x'^2 + y'^2 + z'^2 - c^2t'^2 = x^2 + y^2 + z^2 - c^2t^2$$

and that

$$y' = y \qquad\qquad z' = z$$

as before.

Finally, one can show that one may put

$$x' = a_{11}x + a_{12}t$$
$$t' = a_{21}x + a_{22}t$$

(so that formulas for x' and t' do not involve y and z). Now,

$$x'^2 - c^2t'^2 = (a_{11}x + a_{12}t)^2 - c^2(a_{21}x + a_{22}t)^2 = x^2 - c^2t^2$$

for *all* x and t, and consequently

$$a_{11}^2 - c^2a_{21}^2 = 1$$
$$a_{12}^2 - c^2a_{22}^2 = -c^2$$
$$a_{11}a_{12} - c^2a_{21}a_{22} = 0$$

As the observer in S watches the origin of S' ($x' = 0$, $y' = 0$, $z' = 0$) move he finds $x = vt$; *i.e.*, $x' = 0$ *implies* $x - vt = 0$. Since x' is a linear combination of x and t (*i.e.*, $x' = a_{11}x + a_{12}t$), it follows that

$$x' = \gamma(x - vt)$$

where γ may (and indeed does) depend on v. Thus

$$a_{11} = \gamma \qquad\qquad a_{12} = -\gamma v$$

and one obtains by straightforward manipulations that

$$a_{21} = \pm \frac{1}{c}\sqrt{\gamma^2 - 1}$$
$$a_{22} = \mp \frac{\gamma^2 v}{c}\sqrt{\frac{1}{\gamma^2 - 1}}$$

and

$$\gamma^2 = \frac{1}{1 - v^2/c^2}$$

The formula for t' comes out to be

$$t' = \pm \frac{1}{c}\sqrt{\gamma^2 - 1}\; x \mp \frac{\gamma^2 v}{c}\sqrt{\frac{1}{\gamma^2 - 1}}\, t$$

---

[18] There are various ways of justifying linearity but we shall not go into this here.

and the choice of the signs is determined by the fact that for smaller and smaller v/c one should get closer and closer to the Galilean formula $t' = t$.

Finally, after a few routine simplifications one obtains

$$x' = \frac{1}{\sqrt{1 - (v/c)^2}} (x - vt)$$

$$t' = \frac{1}{\sqrt{1 - (v/c)^2}} \left(t - \frac{v}{c^2} x \right)$$

and this, in its simplest form, is the famous Lorentz transformation. For x' and t' to be real numbers we must have $v < c$. It is a tenet of relativity theory that the relative velocity of observers cannot exceed the speed of light.

One of the most remarkable and striking consequences of the Lorentz formulas is that simultaneity is relative. What observer O would record as simultaneous events (same t, different x's) would not appear so to O'!

Closely related and equally striking consequences are the Lorentz contraction of rods and the time dilatation.

Suppose that observer O' marks off two points $(x_1', 0, 0)$ and $(x_2', 0, 0)$ along his x-axis; he then finds that their distance is

$$l = x_2' - x_1'$$

Observer O who tries to measure this distance could order his helpers to record at a *specified time t* (the helpers, of course, are in possession of *synchronized* clocks) the x-coordinates in their system (S) of the points marked off by O'. The helpers will report to O the numbers

$$x_1 = \sqrt{1 - (v/c)^2}\, x_1' + vt$$

$$x_2 = \sqrt{1 - (v/c)^2}\, x_2' + vt$$

from which O will compute the distance

$$x_2 - x_1 = \sqrt{1 - (v/c)^2}\, (x_2' - x_1') = \sqrt{1 - (v/c)^2}\, l$$

which is shorter than that found by O' by a factor of $\sqrt{1 - (v/c)^2}$.

Similarly, a moving clock will appear to be slower (again by the factor of $\sqrt{1 - (v/c)^2}$) than a clock that is at rest with respect to the observer.

The fundamental importance of Lorentz transformations rests on the principle that *all laws of physics should be invariant under these* transformations.[19] Though this is merely an alternative way of stating the principle of relativity it has the profound effect of making physics analogous to geometry.

For as Euclidean geometry was a study of invariants of the group of linear transformations which leave unchanged the Euclidean distances and angles, so physics is a study of invariants of the Lorentz group that leave invariant the form $x^2 + y^2 + z^2 - c^2 t^2$.

The analogy becomes even more striking if on the one hand one takes plane geometry and on the other hand the two-dimensional (x, t) space-time.

In plane Euclidean geometry rotations of a coordinate system are represented by matrices

$$A = \begin{pmatrix} a_{11} & a_{12} \\ a_{21} & a_{22} \end{pmatrix}$$

such that $A'A = I$, where A' is the transpose of A. It can be shown that all such matrices must be of the form

$$A = A(\theta) = \begin{pmatrix} \cos\theta & \sin\theta \\ -\sin\theta & \cos\theta \end{pmatrix}$$

where θ can be identified with the angle by which the system is rotated.

The matrix defined by the special Lorentz transformation is

$$\begin{pmatrix} \dfrac{1}{\sqrt{1 - (v/c)^2}} & -\dfrac{v}{\sqrt{1 - (v/c)^2}} \\ \dfrac{-v}{c^2\sqrt{1 - (v/c)^2}} & \dfrac{1}{\sqrt{1 - (v/c)^2}} \end{pmatrix}$$

but if instead of x, t and x', t' we take x, ict and x', ict' the transformation assumes the form

$$x' = \frac{1}{\sqrt{1 - (v/c)^2}} \left(x + \frac{iv}{c} ict \right)$$

$$ict' = \frac{1}{\sqrt{1 - (v/c)^2}} \left(-\frac{iv}{c} x + ict \right)$$

---

[19] For the sake of simplicity we consider here only very special Lorentz transformations; *i.e.*, those connecting systems with a common x-axis and with velocity directed along this common axis. They form a very small subgroup of the so-called homogeneous Lorentz group, the group of all linear transformations leaving $x^2 + y^2 + z^2 - c^2 t^2$ invariant *without* the additional requirement that $y' = y$ and $z' = z$. Needless to say, the laws of physics must remain invariant under the larger group.

and the matrix is

$$\begin{pmatrix} \dfrac{1}{\sqrt{1-(v/c)^2}} & \dfrac{iv}{c}\dfrac{1}{\sqrt{1-(v/c)^2}} \\[3mm] -\dfrac{iv}{c}\dfrac{1}{\sqrt{1-(v/c)^2}} & \dfrac{1}{\sqrt{1-(v/c)^2}} \end{pmatrix}$$

It is now possible to find a real θ such that

$$\cos i\theta = \cosh \theta = \frac{1}{\sqrt{1-(v/c)^2}}$$

$$\sin i\theta = -i \sinh i\theta = \frac{iv}{c}\frac{1}{\sqrt{1-(v/c)^2}}$$

and our matrix becomes

$$\begin{pmatrix} \cos i\theta & \sin i\theta \\ -\sin i\theta & \cos i\theta \end{pmatrix}$$

which may be thought of as a rotation through "an imaginary angle" of the "(x, ict)-coordinate system." (The reader should be warned that it is all a manner of speaking, and no mystic significance should be attached to a terminology, especially since the terminology is introduced solely for the purpose of stressing an analogy.)

To see how helpful the geometric view is in formulating physical laws, let us consider briefly the phenomenon of the elastic impact. Let two material points of mass m and M move (in S) along the x-axis with constant velocities u and U respectively, so that an elastic impact results.

To determine the velocities u_1 and U_1 after the impact, one invokes in classical dynamics two laws:

(a) The law of conservation of momentum:

$$mu + MU = mu_1 + MU_1$$

and

(b) The law of conservation of energy:

$$\tfrac{1}{2}mu^2 + \tfrac{1}{2}MU^2 = \tfrac{1}{2}mu_1^2 + \tfrac{1}{2}MU_1^2$$

Before and after the impact the only energy involved is kinetic; thus (b) represents conservation of total energy. Setting $\mu = mM$, we derive by elementary algebra that the solutions of (a) and (b) are either $u_1 = u$, $U_1 = U$, or

$$u_1 = \frac{\mu-1}{\mu+1}u + \frac{2}{\mu+1}U, \qquad U_1 = \frac{2\mu}{\mu+1}u - \frac{\mu-1}{\mu+1}U$$

The first alternative corresponds to the situation when the particles do not collide at all so that only the second alternative is relevant.

The concept of mass is subtle, and there are actually three different concepts of mass. Mass is used

(*a*) as a measure of the "amount of matter,"
(*b*) as a measure of "resistance to changes in motion,"
(*c*) as a gravitational "charge."

In connection with (*a*) one speaks of the *proper mass*, with (*b*) of the *inertial mass*, and with (*c*) of the *gravitational mass*.

In classical dynamics no distinction among the three is made; for the time being we shall simply assume that there is some method whereby a piece of matter can be assigned a number (in some units) called its *proper mass* and that the proper masses enter the laws of conservation of momentum and energy above. We shall also assume that the proper mass is conserved in all physical processes.

We can now see that neither conservation of momentum nor conservation of energy in the forms stated above can be a law of physics. For if the impact process is observed by O' from the system S', this observer will find that before collision the velocities were

$$u' = \frac{u - v}{1 - \dfrac{uv}{c^2}} \qquad U' = \frac{U - v}{1 - \dfrac{Uv}{c^2}}$$

while after collision they are

$$u_1' = \frac{u_1 - v}{1 - \dfrac{uv}{c^2}} \qquad U_1' = \frac{U_1 - v}{1 - \dfrac{U_1v}{c^2}}$$

To see how, *e.g.*, one gets the formula for u' note that in S, $u = \Delta x / \Delta t$ (displacement Δx divided by the time Δt during which the displacement occurred). In S', $u' = \Delta x' / \Delta t'$ where $\Delta x'$ and $\Delta t'$ are obtained from Δx and Δt at the Lorentz transformation. Thus

$$u' = \frac{\Delta x - v\Delta t}{-\dfrac{v}{c^2}\Delta x + \Delta t} = \frac{u - v}{1 - \dfrac{uv}{c^2}}$$

It is now a simple matter to check that

$$mu' + MU' \neq \tfrac{1}{2}mu_1' + \tfrac{1}{2}MU_1'$$

and

$$\tfrac{1}{2}mu'^2 + \tfrac{1}{2}MU'^2 \neq \tfrac{1}{2}mu_1'^2 + \tfrac{1}{2}MU_1'^2$$

Let us now see how the concept of invariance can guide us toward an appropriate modification of the law of conservation of momentum.

In classical physics the velocity of a moving particle is a vectorial quantity describable in a rectilinear coordinate system by its x,y,z components, u_x, u_y, u_z.

Relativity theory takes the view that the world is inherently four-dimensional with time playing the role of the fourth dimension; on the other hand, it also takes the view that spatial and temporal dimensions are forever intermingled through the Lorentz transformation. As a consequence vectorial quantities in relativity theory must be "four-vectors;" *i.e.*, objects describable in each frame by four components

$$w_x, \; w_y, \; w_z, \; w_t$$

that transform from frame to frame by the Lorentz transformations in the same way the coordinates x,y,z,t transform.

In other words, in S' the components of the vector w are

$$w_{x'} = \frac{w_x - v w_t}{\sqrt{1 - (v/c)^2}}$$

$$w_{y'} = w_y \qquad w_{z'} = w_z$$

$$w_{t'} = \frac{-\dfrac{v}{c^2} w_x + w_t}{\sqrt{1 - (v/c)^2}}$$

It should be clear now that there is no way to augment u_x, u_y, u_z, by a fourth component u_t to produce a four-vector. In fact, if (u_x, u_y, u_z, u_t) were components of a four-vector in S', then the y-component in S' would have to be

$$u_{y'} = u_y$$

On the other hand u_y' must also be the y-component of ordinary (classical) velocity as measured from S' and hence

$$u'_y = \frac{\sqrt{1 - (v/c)^2}}{1 - uv/c^2} u_y$$

in contradiction to $u_y' = u_y$.

This follows from observing that $u_y = \Delta y / \Delta t$, where Δy is the displacement in the y-direction that takes place during the time interval Δt. Similarly,

$$u_{y'} = \frac{\Delta y'}{\Delta t'} = \frac{\sqrt{1 - \left(\frac{v}{c}\right)^2}\Delta y}{(\Delta t - \frac{v}{c^2}\Delta x)} = \frac{\sqrt{1 - \left(\frac{v}{c}\right)^2}u_y}{1 - \frac{vu}{c^2}}$$

as noted above.

On the other hand it is easily checked that the four numbers

$$\frac{u_x}{\sqrt{1 - (u/c)^2}}, \qquad \frac{u_y}{\sqrt{1 - (u/c)^2}}, \qquad \frac{u_z}{\sqrt{1 - (u/c)^2}}, \qquad \frac{c}{\sqrt{1 - (u/c)^2}}$$

where $u^2 = u_x^2 + u_y^2 + u_z^2$ is the square of the classical speed, are components of a four-vector. (In guessing the above form of the four-velocity we are guided by the fact that for every four-vector w, $w_x^2 + w_y^2 + w_z^2 - c^2 w_t^2$ is the same in all Lorentz frames.) Moreover, for speeds u that are small compared to the speed of light c, the spatial components of our four-vector reduce to components u_x, u_y, u_z of ordinary velocity.

We now can define the four-momentum as the proper mass m times the four-velocity above; *i.e.*, the components of four-momentum in S are

$$\frac{mu_x}{\sqrt{1 - (u/c)^2}}, \qquad \frac{mu_y}{\sqrt{1 - (u/c)^2}}, \qquad \frac{mu_z}{\sqrt{1 - (u/c)^2}}, \qquad \frac{mc}{\sqrt{1 - (u/c)^2}}$$

One is tempted to hope that the classical law of conservation of momentum and energy will be replaced by the conservation of the relativistic four-momentum.

Experiment, the eventual test of all physical theories, has amply confirmed this relativistic law, but what concerns us here most is the remarkable way in which geometric considerations point the way toward deep descriptions of physical reality.

How deep our newly found law of conservation of four-momentum is can be gauged from the conservation of the fourth component.

For small u/c we have (approximately)

$$\frac{mc}{\sqrt{1 - (u/c)^2}} \sim mc(1 + \frac{1}{2}\frac{u^2}{c^2}) = \frac{1}{c}(mc^2 + \frac{1}{2}mu^2)$$

and we recognize $\frac{1}{2}mu^2$ as the classical kinetic energy. In classical treatment of impact, conservation of energy was quite separate from conservation of momentum. Now, the two conservation laws are merged by geometry into one! But this is not all. In the formula

$$\frac{mc}{\sqrt{1-(u/c)^2}} \sim \frac{1}{c}(mc^2 + \frac{1}{2}mu^2)$$

we see, in addition to the familiar classical kinetic energy $\frac{1}{2}mu^2$, the wholly new energy term mc^2. This is the famous rest energy of Einstein and, although a great deal of physical interpretation is needed to fully appreciate its meaning and significance, one cannot escape the wonder that its presence in nature could be revealed by geometry.

15. *Transformations, Flows, and Ergodicity*

Suppose now that E is the Euclidean space and T is a one-to-one transformation of the space onto itself; *i.e.*, into the whole of the space. Starting with a point p we consider the points $T^{-n}(p)$, $T^{-n+1}(p)$, ... p, $T(p)$, $T^2(p)$, ..., $T^n(p)$, ... One is properly interested in the behaviour of this (finite or infinite) sequence of points. The study of such sequences is the core of the theory of ergodic properties of transformations. We shall briefly describe, by examples, some of the problems that arise in this context.

Suppose E is the unit cube or the unit sphere in 3-space (three-dimensional space) and imagine that it is filled with an incompressible fluid that moves in a steady flow throughout E. The flow is called steady if at each point of space the direction and speed of motion do not depend on time. Given such a flow, we may consider a transformation $T(p)$, where p is any point of the space, defined as follows: $T(p)$ is the position of the fluid particle at p, one second later. To obtain the position of the original particle after two seconds one merely has to look at $T^2(p)$ and, in general, after n seconds, at $T^n(p)$. ("One second" is, of course, an arbitrary unit of time.) The flow is assumed to be incompressible. By this we mean that if we start with an arbitrary subregion A of E, with volume $m(A)$, and look at the set of points occupied by the particles that were originally in A one time-unit later, then the volume of $T(A)$ will be the same. In other words $m(A) = m(T(A))$ for all A.

Such flows are studied in hydrodynamics, but they are also important in the general study of dynamic systems. This arises from the following way of looking at problems in mechanics. Given a dynamic system consisting of n material points in 3-space and given their initial position and momenta vectors, we may represent the full system by one point in a $6n$-dimensional space. (We need 3 spatial coordinates and 3 momenta components for each of the n points.) The space of these points is called phase space of our system. The magnitudes and directions of forces between the points (they may depend, *e.g.*, only on their mutual distance) are mathematically prescribed. As time goes on, positions and momenta will change according to the equations of dynamics, and each representative point will move through phase space. We thus have a flow in the $6n$-dimensional space. Liouville proved that for conservative dynamic systems the flow is incompressible; *i.e.*, it preserves the volumes in phase space.

The term conservative means that energy is conserved, and this in turn means that a certain function of positions and momenta remains constant throughout the motion. Thus the representative point is constrained to move on a surface E of constant energy (called the energy surface), and the evolution of the dynamic system defines a flow on this surface.

Preservation of ordinary volume by the flow in the full phase space *induces* a preservation of a specific well-defined (but somewhat complicated) measure on the energy surface. Thus for a large collection of subsets A of the energy surface E there is a countably additive measure m such that

$$m(A) = m(T(A))$$

If we observe the representative point after $1, 2, \ldots, n, \ldots$ seconds, we are dealing with the iterates of the transformation T that preserves the measure m on E.

A fundamental hypothesis, first formulated by Boltzmann, states that in course of time and "in general" (*i.e.*, for most dynamic systems) a trajectory of the representative point will pass through all the points of the energy surface. This was the original "ergodic hypothesis." It was soon realized that this is impossible, and the impossibility was demonstrated on purely topological grounds: a "curve" (*i.e.*, a one-to-one continuous image of an infinite line) cannot pass through all points of an energy surface if $n > 2$. The original postulate was unnecessarily strong; for the purposes of Boltzmann's statistical mechanics, a weaker property would suffice. In particular, for such a "general transformation" the curve would have to pass arbitrarily near any given point of the energy surface E. In other words, the sequence of points $p, T(p), T^2(p), \ldots,$ $T^n(p), \ldots$ would have to be *dense* in this set. One would like to know if this is the case for many or "most" transformations T that *preserve volume*.

The existence of transformations T that are volume-preserving and are "ergodic" in the above sense has been established. It even has been shown that transformations without this property are in a well-defined sense exceptional among all measure-preserving transformations. Actually, even more can be shown: the sequence of iterates of a point is not merely dense in space, but it is *uniformly* dense. This asymptotically uniform behaviour of the sequence of points $p, T(p), \ldots, T^n(p), \ldots$ can be defined as follows: consider a set A with measure $m(A)$. Starting with a point p at time 0, one asks for the frequency with which the iterates of p fall into A. This frequency of "hits" of A can be written down concisely in symbols. We define the function $\chi_A(p)$, the characteristic function of the set A, by putting $\chi_A(p) = 1$ if p belongs to A, and $\chi_A(p) = 0$ otherwise. The expression

$$\frac{1}{N} \sum_{i=1}^{N} \chi_A(T^i(p))$$

is the desired frequency for iterates from 1 to N. The ergodic theorem asserts that in the limit $N \to \infty$ and for almost every point p ("almost every" in the sense of the measure m), this frequency will be equal to the relative measure of the region A or

$$\lim_{N \to \infty} \frac{1}{N} \sum_{i=1}^{N} \chi_A(T^i(p)) = \frac{m(A)}{m(E)}$$

G. D. Birkhoff first proved that the limit in question *exists* for almost all p. It was later shown that there exist many transformations for which this limit is indeed equal to $m(A)/m(E)$. In fact one could prove that in a specific sense *most* volume-preserving transformations possess this property.

If the transformation (of a finite set of integers into itself, or a continuous transformation of the interval into itself, or of the n-dimensional space into itself) is not one-to-one, the process of iteration can still be performed. It is much harder to determine in such a case the properties of the sequence of iterated images of individual points. In some isolated instances it is still possible to say something about their behaviour. Take, for example, a transformation of the interval $(0,1)$ on itself, defined by $x' = f(x) = 4x(1 - x)$. By iterating this function whose graph is a parabola, one obtains functions that are polynomials of increasingly higher orders and whose graphs will show an increasing number of maxima and minima. In this case it is possible to prove that, starting with almost every point, the sequence of iterated images will be dense in the whole interval.

Because of an accidental feature of the transformation, even more can be proved. If we set $x = \sin^2\theta$, we obtain $f(x) = 4 \sin^2\theta \ (1 - \sin^2\theta) = \sin^2 2\theta$. Hence the transformation x into $4x(1 - x)$ is equivalent to the transformation θ into 2θ which is much easier to study.

In general, however, it is very difficult to determine properties of iterates, even of simple algebraic transformations.

It should be mentioned that an extension of the ergodic theorem to transformations that are not one-to-one is possible as long as one defines the term "measure-preserving" to mean that the measure of the "inverse image" of a set A (*i.e.*, the set that is transformed into A) has the same measure as A.

A striking application of the ergodic theorem for transformations that are not one-to-one is to the theory of continued fractions. Let x be a real number between 0 and 1. To represent x as a continued fraction we proceed as follows: Take the inverse of x and write it as an integer a_1 closest to it plus a nonnegative fraction x_1. Now do to x_1 what was done to x and repeat the process indefinitely, or until it terminates, in case x is a rational number.

For example,

$$\frac{17}{21} = \frac{1}{\dfrac{21}{17}} = \frac{1}{1 + \dfrac{4}{17}} = \frac{1}{1 + \dfrac{1}{\dfrac{17}{4}}} = \frac{1}{1 + \dfrac{1}{4 + \dfrac{1}{4}}}$$

or

$$\sqrt{2} - 1 = \frac{1}{\sqrt{2} + 1} = \frac{1}{2 + (\sqrt{2} - 1)} = \frac{1}{2 + \dfrac{1}{\sqrt{2} + 1}} =$$

$$\frac{1}{2 + \dfrac{1}{2 + \dfrac{1}{2 + \dfrac{1}{2 + \cdots}}}} = \cdots$$

The algorithm of expanding x in a continued fraction is summarized as follows

$$\frac{1}{x} = a_1 + x_1, \qquad \frac{1}{x_1} = a_2 + x_2, \text{ etc.}$$

so that finally

$$x = \frac{1}{a_1 + \dfrac{1}{a_2 + \dfrac{1}{a_3 + \cdots}}}$$

If we now define the transformation Tx of the interval $(0,1)$ onto itself by the formula

$$Tx = \frac{1}{x} - \left[\frac{1}{x}\right] \qquad \left[\frac{1}{x}\right] = \text{integer closest to but not exceeding } 1/x$$

we have

$$a_1(x) = \left[\frac{1}{x}\right] \qquad a_2(x) = a_1(Tx), \qquad a_3(x) = a_1(T^2x), \ldots$$

and we are dealing again with an iteration of a transformation.

The inverse image of the interval (a,b), where $0 < a < b < 1$, is the infinite union of all intervals

$$\left(\frac{1}{1+b}, \frac{1}{1+a}\right), \left(\frac{1}{2+b}, \frac{1}{2+a}\right), \left(\frac{1}{3+b}, \frac{1}{3+a}\right), \ldots$$

If we define the measure of an interval (α, β) to be

$$\frac{1}{\log 2} \log \frac{1+\beta}{1+\alpha} = \frac{1}{\log 2} \int_{\alpha}^{\beta} \frac{dx}{1+x}$$

we can check that this measure of (a,b) is equal to the sum of the measures of the intervals composing the inverse image of (a,b); from this it follows that if we define the measure m of a subset A of the interval $(0,1)$ by the formula

$$m(A) = \frac{1}{\log 2} \int_A \frac{dx}{1 + x}$$

This measure is preserved by the transformation T. One can then show that the ergodic theorem is applicable; for example, one obtains the following result:

For almost every x (in the sense of the measure m or, equivalently, in the sense of ordinary Lebesgue measure) the frequency with which the integer k appears in the sequence $a_1, a_2, \ldots$ is

$$\frac{1}{\log 2} \log\left(\frac{(k + 1)^2}{k(k + 2)}\right)$$

This is obtained from our statement of the ergodic theorem by taking E to be the interval $(0,1)$; A to be the interval $(1 + \frac{1}{k + 1}, 1 + \frac{1}{k})$, that is, the set on which $a_1(x) = k$; and the measure m to be the logarithmic measure defined above.

If we have dwelt so much on this example, it was to illustrate once again the miracle of a theory originating in one context and then emerging in a decisive way in a totally unrelated one. In our case, the ergodic theory that originated in Boltzmann's attempts to place kinetic theory on a solid ground was applied to the theory of continued fractions.

16. *More on Iteration and Composition of Transformations*

The simplest transformations of the n-dimensional Euclidean space into itself are linear transformations.

As we have seen in Section 13, such a transformation is of the general form

$$\begin{aligned}
x_1' &= t_{11}x_1 + t_{12}x_2 + \cdots t_{1n}x_n \\
x_2' &= t_{21}x_1 + t_{22}x_2 + \cdots t_{2n}x_n \\
x_3' &= t_{31}x_1 + t_{32}x_2 + \cdots t_{3n}x_n \\
&\cdots\cdots\cdots\cdots\cdots\cdots\cdots\cdots\cdots\cdots \\
x_n' &= t_{n1}x_1 + t_{n2}x_2 + \cdots t_{nn}x_n
\end{aligned}$$

where the t_{ij} entries are real numbers. As before, the matrix of coefficients will be denoted T. An important and interesting case is where all the t_{ij}'s are positive or rather nonnegative.[20] Such matrices occur in many applications in algebra, in probability theory, in the study of distribution of particles in multiplicative systems like those of neutrons in reactors, and in mathematical models in economics. If T is such a positive matrix, one verifies immediately that vectors located in a positive "octant" of space go under T again into such a set. By a positive "octant" we mean the set of all vectors whose components $x_1, x_2, \ldots, x_n$ are positive.

---

[20] The positivity of the elements of T is not an intrinsic property of the linear transformation. In a different coordinate system the matrix will generally lose this property. It is possible though to define a "positive transformation" intrinsically as one that transforms a specific unbounded convex set (called an "octant") into itself. One can then show that there is a coordinate system in which the transformation is represented by a matrix with nonnegative elements.

Consider now the surface of the unit sphere in our space and in particular the part of it in the positive octant; *i.e.*, the set C of all vectors $(x_1, x_2, \ldots, x_n)$ whose components are positive and satisfy the equation $x_1^2 + x_2^2 + \cdots + x_n^2 = 1$. We shall define another transformation S as follows: Given any vector x in the positive octant, take its multiple which lies on the unit sphere and call it x^*. We define $S(x^*)$ to be equal to $(T(x))^*$. S transforms that part of the unit sphere that is in the positive octant into itself. This set of points is topologically equivalent to the $(n-1)$-dimensional cube. In two dimensions it would be an arc of the circle, topologically the same as an interval. In three dimensions it would be an octant of the space of the sphere, topologically the same as a disk, and so on. We have stated previously that a continuous transformation of the n-dimensional cube or sphere into itself must possess a fixed point (Brouwer's Fixed-point Theorem). From this we conclude that there must exist a point on the unit sphere x_0^* such that $S(x_0^*) = x_0^*$. In view of the definition of S, it follows that there must exist a vector x_0 in the positive octant that goes into some multiple of itself under T: $T(x_0) = \lambda x_0$. We recognize x_0 as an eigenvector and λ as the corresponding eigenvalue of T. Although the proof is more elaborate, one also can prove that this vector is unique for matrices whose entries are strictly positive and that, if one starts with any vector x in the positive region of space and forms a sequence $x, T(x), T^2(x), \ldots, T^n(x), \ldots$, these vectors will converge *in direction* to this unique eigenvector. This theorem, first proved by Frobenius, has many interesting applications.

Matrices with nonnegative elements are used, perhaps most extensively, in the theory of Markov chains. Markov chains are a generalization of the concept of independent trials.

It often happens that a system that can be found in any one of the states s_1, s_2, s_3, ... can make transitions from one state to another and that these transitions are governed by chance. We can assume that it takes time τ to complete a transition. With the transition from s_i to s_j there is associated a probability (basic or one-step transition probability)

$$p_{ij} = \text{Prob. } \{s_i \rightarrow s_j \text{ in time } \tau\}$$

and it is assumed that the probability associated with a sequence of n simple transitions (each taking time τ to complete)

$$s_i \rightarrow s_{i_1} \rightarrow s_{i_2} \rightarrow \ldots \rightarrow s_{i_{n-1}} \rightarrow s_j$$

is

$$p_{ii_1} p_{i_1 i_2} \cdots p_{i_{n-1} j}$$

This "product assumption" is at the heart of the concept of a Markov chain.

To give an example of a Markov chain, let us consider two boxes I and II, into which we somehow distribute $2R$ numbered balls. Every τ seconds we choose a number from 1 to $2R$ "at random" (*i.e.*, each number with probability $1/2R$ with successive choices being independent of all others) and move the ball so numbered from its present box to the other.

We can think now of the number i of balls in box I as defining the state of the system, and we see that the only transitions possible are $i \rightarrow i - 1$ or $i \rightarrow i + 1$, except when $i = 0$, in which case only $0 \to 1$ is possible, or when $i = 2R$, in which case we must have $2R \to 2R - 1$.

The one-step transition probabilities are clearly:

$$p_{ij} = 0 \ (\text{if } j \neq i - 1, i + 1)$$
$$p_{i,i-1} = \frac{i}{2R}$$
$$p_{i,i+1} = 1 - \frac{i}{2R}$$

and the "product assumption" follows from the independence of the successive drawings (choices) of numbers from 1 to $2R$.

The model just described was introduced by Paul and Tatiana Ehrenfest in 1907 to illustrate some logical difficulties that occur when one attempts to reconcile time-reversible laws of dynamics with the irreversibility of thermal processes, as demanded by the second law of thermodynamics.

We shall come back to this model in Chapter 3, but for the time being we continue with the general discussion of Markov chains.

From the product assumption and the axiom of additivity it follows that the probability that the system, originally in s_i, will be found in state s_j after n transitions (*i.e.*, after time $n\tau$) is the (i,j) element of the nth power of the matrix P of the transition probabilities. The desired probability is the sum over *all* $i_1, i_2, \ldots, i_{n-1}$ of

$$p_{ii_1} p_{i_1 i_2} \cdots p_{i_{n-1} i}$$

and this is precisely the (i,j) element of P^n. Thus,

Prob.$\{ s_i \to s_j$ in time $n\tau \} = (i, j)$ element *of* P^n

where

$$P = \begin{pmatrix} p_{11} & p_{12} & \ldots \\ p_{21} & p_{22} & \ldots \\ \ldots & \ldots & \ldots \end{pmatrix}$$

It is important to recall that matrix multiplication (and in particular raising a matrix to a power) came up directly in connection with composition of linear transformations and iteration of a linear transformation in particular.

Here, however, we encounter the algebraic operation of raising a transformation to the power n in a context that has nothing to do with the motivation that originally led us to introduce and to study this operation.

But once such a miracle happens (and it happens more often than one has any right to expect), it can be seized upon and exploited.

One thus can apply the extensive knowledge of matrices gained from algebra and geometry to the study of Markov chains.

Conversely, since examples of Markov chains often arise outside mathematics (*e.g.*, in physics), one is provided with new sources of intuition in dealing with problems concerning iteration of matrices. Through the thread of a common theme (*e.g.*, iteration), far-removed areas of thought are brought close together with untold mutual benefits.

Iteration is the simplest illustration of composing transformations; in an iteration we always compose the transformation or function with itself. Suppose we have two transformations (or functions) S and T of a set E onto itself to start with and, instead of a group generated by iteration, we consider all possible products of S and T; *i.e.*, we consider T^2, S^2, $T(S)$, $S(T)$, $S^2(T)$, STS, TST, etc., also the inverses T^{-2}, S^{-2}, . . . , and all combinations such as $T(S^{-1})$ or $S(T^{-1})$ or $T(S^{-1}(T))$. In short, we consider the full group generated by two transformations. The number of elements in this group is still countable, but the group itself may be very extensive and rich in the forms of transformations that are its members.

It can be proved that, starting with two one-to-one continuous transformations T and S of the interval into itself, one will obtain by compositions of these two a whole dense class of transformations of the interval into itself. In other words, given an arbitrary continuous transformation R and a number $\epsilon > 0$, one can find a product transformation P composed of a finite number of T, S, T^{-1}, and S^{-1} such that $|P(x) - R(x)| < \epsilon$ for all x; *i.e.*, any continuous transformation can be approximated arbitrarily closely by transformations of our class.

An analogous theorem holds in higher-dimensional spaces: if E is the n-dimensional sphere, one can find a finite number (actually, four will suffice) of homeomorphic transformations of E onto itself (homeomorphism means one-to-one continuous transformation onto itself) such that their compositions permit one to approximate arbitrarily closely any given homeomorphism. The proof is such that it is hard to say what the precise properties of these homeomorphisms are. It would be useful if one could show, for example, that these transformations could be chosen from the class of those that are everywhere differentiable. If this were the case, one would settle one of the outstanding problems in topology: Is every general homeomorphism of the n-dimensional space approximable by differentiable homeomorphisms?

We should stress here that our problem of approximability concerns homeomorphisms; *i.e.*, one-to-one continuous transformations. If we do not require the one-to-one character, the answer is affirmative, since every transformation can be approximated by another that is differentiable by using the theorem of Weierstrass: every continuous function on a bounded region in n-space can be approximated by polynomial functions.

A few more problems concerning the operation of composition will show how quickly one reaches the boundary of the unknown. Suppose E is the Euclidean plane, and we consider the group of all homeomorphic transformations obtained by composing homeomorphisms of the form

$$\begin{array}{cc} x' = f(x,y) & x' = x \\ & \text{or} \\ y' = y & y' = g(x,y) \end{array}$$

It turns out that an arbitrary homeomorphism given by $x' = \phi(x,y)$ and $y' = \psi(x,y)$ can be approximated by those from our group above. In three or more dimensions analogous problems are still open. For example, in three dimensions we may permit generating transformations of the form $x' = f(x,y,z)$, $y' = y$, $z' = z$ and its two analogues, or let the group G be generated by all homeomorphisms of the type $x' = f(y,z)$, $y' = g(x,z)$, $z' = h(x,y)$. In both cases the question whether arbitrary homeomorphisms of the n-dimensional space are approximable by transformations of the above type is still open.

Finally, we might mention the more recent beautiful results of Kolmogorov and Arnold on representation of arbitrary real-valued continuous functions of any number of variables as compositions of such functions of only two variables. It turns out that continuous functions of many variables can be represented not just approximately, but *exactly*, as compositions of only a finite number of functions of two variables; *e.g.*,

$$f(x_1,x_2,x_3,x_4) = h_1[h_2(x_1,x_2),h_3(x_3,x_4)]$$
$$\text{or}$$
$$f = h_1\{h_2[h_3(x_1,x_2),h_4(x_3,x_4),h_5(x_1,x_4)]\}$$

and so on.

17. *Proving the Obvious*

In this section we shall elaborate a bit on the relations between elementary intuition and mathematical rigour. Our examples will be taken mainly from topology and will be attempts to illustrate how our geometrical intuition allows us in certain cases to state mathematical facts that are provable only by rather elaborate and sometimes very difficult methods. On the other hand, there are intuitively plausible statements that are *not* valid, but examples disproving them can be found only with difficulty.

One of the fundamental theorems in the topology of the Euclidean plane states that a simple closed curve divides the plane into two regions, one inside the curve and the other outside. A simple closed curve is a set of points on the plane that is a one-to-one continuous image of the circumference of a circle. By this we mean that a continuous transformation is defined on the circumference of the circle and is such that two different points on the circumference correspond to two different points of the plane. The set of the image points then has the property of dividing the plane; *i.e.*, there will be two disjoint sets of points on the plane that cannot be connected by an arc without crossing this set (curve). The statement seems entirely obvious; a rigorous proof, however, is not very simple and the first one was obtained by Camille Jordan. Indeed, if the curve is very contorted (fig. 24) or takes many turns so that it resembles a tightly wound spiral, it is visually difficult to distinguish the inside from the outside. (There is a game that used to be played at county fairs as follows: A person winds a doubly folded string or a leather belt and puts the ends of it together. An onlooker has to guess, by sticking a pencil between two contiguous pieces of the curve, whether it will be on the inside or on the outside of the belt when it is unwound. Evidently a dishonest demonstrator can always win this game by putting the ends together in one fashion or another.)

An object like "a simple closed curve" embodies an idea not as clear in its full generality as the examples that give rise to it, for most of the examples are of convex or only moderately convoluted curves. We shall return to this in Chapter 2.

Another theorem follows that may seem obvious and yet is not at all easy

Fig. 24. Which points of the plane are inside and which are outside this curve?

to prove. Try to imagine on the surface of a sphere a continuous distribution of vectors. That is, try to attach to every point on this sphere a short segment of fixed length tangent to the sphere and whose direction varies continuously as one moves on the sphere. This turns out to be impossible! A sphere cannot be "combed." There will always be a point where the direction of the vector cannot be tangent to this sphere, a sort of whirlpool point. This impossibility follows from a topological theorem of Brouwer: for every continuous transformation of the sphere into itself either there must exist a pair of fixed points or else some point must be transformed into the antipode of itself. Again, the impossibility of "combing" a sphere might be completely obvious intuitively and yet no simple complete proof of this fact is known.

Consider another theorem about the surface of the sphere. Suppose that to each point on the surface of the sphere there are attached two real numbers that vary continuously on the sphere. In other words, we have two real-valued continuous functions $f_1(p)$ and $f_2(p)$ defined on the surface of the sphere. There must then exist at least one p_0 such that at this point and at its antipode, denoted p_0^*, both functions assume the same values. In other words,

$$f_1(p_0) = f_1(p_0^*)$$
$$f_2(p_0) = f_2(p_0^*)$$

For example, consider the temperature and the pressure as the two functions defined for every point on the surface of the earth (assumed spherical). If we assume that these functions vary continuously, there must be a point on the earth where the temperature is the same as the temperature at its antipodal point; the pressure at this point is the same as the pressure at its antipode. This theorem has some amusing consequences, one known as the "ham sandwich" theorem. Given any three solids in space, there exists a single plane that will cut each of the solids into equal volumes. (The solids could be a piece of bread, ham, and butter in any position whatever, hence the name.) We will briefly sketch the proof of this statement based on the theorem on antipodal points (whose proof is not simple). Parallel to every direction of space, represented as a line through the centre of a fixed sphere, there will be a plane dividing the first of the three volumes into two equal parts. This can be deduced from continuity of the two volumes determined by a single plane. Now let us see what this plane does to the other two solids. Denote $f_1(p)$ the difference in the two volumes into which this plane cuts the second set. The set might, of course, be found only on one side of this plane; in that case the difference would be the entire volume of the set taken with its proper sign. We will have analogously a second function $f_2(p)$ for the second volume. These two functions will be defined for every point on the surface of the sphere. From our theorem on antipodes it follows that there must exist a point and its antipode where the functions have equal value. The sign is reversed by passing to the antipode; this follows from our definition of the functions and the only number that is equal to its negative is 0, and our assertion follows.

Chapter 2   Themes, Trends, and Syntheses

PERHAPS THE MOST striking feature of mathematics as an intellectual discipline is the enormous variety of the problems with which it deals. If one thinks, for example, of the problem of the number of ways of changing a dollar and of the problem of constructing $\sqrt[3]{2}$ using the ruler and compass, seemingly so unconnected, one cannot fail to marvel at their being meaningfully related.

This variety, coupled with a lack of clear-cut criteria for what forms the subject matter of mathematics, makes large-scale syntheses and unifications extremely difficult to achieve. One must also beware of unifications of such exalted generality as to be trivial and avoid syntheses so rigid that they constrain future growth and developments. It should be noted that the only serious attempt in recent years to present the whole of mathematics from a unified point of view, that of the Bourbaki group, was criticized on *both* of these grounds.[1]

Before the middle of the 19th century there was little conscious effort at synthesis or unification. Of course, Euclid's *Elements* represent a major synthesis and Descartes' analytic geometry was a major unification of algebra and geometry, but mathematicians after Newton were too busy joyfully exploring the new vistas opened by the algorithms of calculus to take time off to organize their rapidly expanding realm.

Then came a reaction, and a trend towards organization developed that still is continuing.

One reason for the change was that the body of mathematics had grown so large that some organization was becoming necessary lest parts of the subject cease to communicate with one another. Also, unrestrained intuition, unhampered by the rigid standards imposed by a formal system, was beginning to get mathematicians into trouble.

Euclid once was the unsurpassed model of rigour. But as mathematicians were exposed to an ever widening stream of problems, critical senses sharpened and logical senses grew subtle and more refined.

To an 18th-century mathematician the fact that a closed simple curve cuts the plane into two parts was so obvious as to be hardly worth mentioning. But in the 19th century Jordan, who understood the subtlety of the problem, attempted without full success to provide a proof. (Even today there is no really simple proof of the Jordan Curve Theorem.)

[1] " 'Bourbaki, Nicholas,' " *Encyclopædia Britannica* (1968).

A mathematician of the 18th or early 19th century used the concept of a simple closed curve in a purely intuitive way. By Jordan's time, a simple closed curve was understood to be the result of a *continuous one-to-one mapping of a circle*. By a circle we mean the circumference of a circle; if we include the interior, we speak of a disk. Since a circle clearly cuts the plane into its interior and its exterior, it seems obvious that a continuous one-to-one image of a circle will preserve this property. Perhaps we think this is so obvious because we endow continuity with all sorts of properties. Since our intuition is both inspired and limited by physical reality, a "simple, closed curve" invokes an image of a relatively smooth curve with perhaps a number of sharp corners; something like fig. 25 at worst.

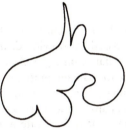

Fig. 25

But "continuous one-to-one image of a circle" (which is the *formal* definition of a simple closed curve) can determine quite a "wild" set. For example, such a curve may be of infinite length and, worse yet, it may be such that at *none of its points* can one draw a tangent to it!

The first example of such a curve (given by Weierstrass) has the following parametric equations:

$$x = \sin \theta$$
$$y = \sum_{n=1}^{\infty} \frac{1}{2^n} \cos 3^n\theta$$

It is clear that x is a continuous function of θ. Since the coefficients $\frac{1}{2^n}$ of the infinite series that defines y decrease geometrically and since $\cos 3^n\theta$ always lies between -1 and $+1$, it may be shown that y, too, is a continuous function of θ. Thus as θ changes from 0 to 2π the corresponding point $(x(\theta), y(\theta))$ moves *continuously* along a specific curve. For the curve to have a tangent at a point corresponding to θ both derivatives $dx/d\theta$ and $dy/d\theta$ must exist.

Weierstrass proved that $dy/d\theta$ fails to exist for all θ. Although this is not easily proved, one is readily led to suspect that this could be so by noting that formal differentiation of the series for y (term by term) leads to the series

$$\sum_{n=1}^{\infty} \left(\frac{3}{2}\right)^n \sin 3^n\theta$$

which *diverges* for all but a countable set of θ.

If the "one-to-one" requirement is dropped, one can construct (as Peano first did) a continuous image of an interval (*i.e.*, a curve, though not a simple one) which fills a square. If one thinks of such "pathological" creations that are allowed if only the mere continuity is postulated, one begins to understand Jordan's need to prove his theorem and that the proof is quite far from being simple.

For "smooth" curves, such that a tangent can be drawn at "nearly" every point (with at most a finite number of exceptions), the Jordan Curve Theorem is much easier to prove.

Not until about 1830 were continuity, smoothness, and related concepts defined with some clarity; before then they had been used informally and sometimes haphazardly.

Recall that the middle of the 19th century marked the beginning of a new era in mathematics, characterized by growing distrust of intuition not backed by proofs and by increased reliance on logic. As a result, mathematics tended to appear more austere, more formal, and more inwardly oriented. Nothing was taken for granted and nothing escaped close scrutiny. Even Euclid was subjected to an exhaustive logical analysis and cracks appeared in his magnificent edifice. For example, Euclid neglected to state a whole group of axioms that are needed to formalize the concept of "betweenness," the so-called axioms of order. These axioms appear so obviously and so trivially "true" that Euclid and his followers took them for granted. But complete formalization should mean that geometry may be taught to a blind man and even to a computer. Many of Euclid's arguments depend on using the fact that a point D on a line determined by points A, B lies between these points. The familiar proof that a triangle ABC with $AC = BC$ is such that $\sphericalangle A = \sphericalangle B$ calls for dropping a perpendicular from C on AB; this perpendicular intersects AB at D and to complete the proof one needs the fact that D is *between* A and B. There is no way to "explain" this to a blind man or to a computer without formalizing the notion of lying in-beween. More complete axiomatization of geometry was accomplished in 1899 by Hilbert in his famous *Grundlagen der Geometrie*.[2]

The concept of number also was subjected to careful analysis that stimulated the growth of both a new algebra and logic itself.

In fact, mathematical algebra became what it is largely today: a study of such abstract systems as groups, rings, and fields. Consider the following example of what the "spirit" of algebra is and how it can pervade branches of mathematics that traditionally have been separated from algebra.

---

[2] Even Hilbert's axiomatization may be a cause for concern to the strict logician because of the axiom of continuity. This axiom, which in effect establishes a one-to-one order-preserving correspondence between points on a straight line and real numbers, transfers to geometry all the difficulties connected with the nondenumerability of the set of real numbers.

Let us begin by assuming that we know what integers (positive, negative, and 0) are.

Integers can be added, subtracted (*i.e.*, they can be solutions of equations like $a + x = b$), and multiplied. These operations have the following properties:

1. $a + b = b + a$
2. $a + (b + c) = (a + b) + c$
3. $ab = ba$
4. $(ab)c = a(bc)$
5. There is one and only one integer, namely 0, such that $a + 0 = a$ for every integer a.
6. There is one and only one integer, namely 1, such that $a \cdot 1 = a$ for every integer a.
7. There are no divisors of 0; *i.e.*, $ab = 0$ implies that either $a = 0$, or $b = 0$, (or both are 0).

One proceeds as follows to construct the class of rational numbers.

Consider all ordered pairs of integers in which the second member is not 0 and call two pairs (a, b) and (c, d) *equivalent*, in symbols $(a, b) \sim (c, d)$, when $ad = bc$.

Thus defined, equivalence has the following properties:

(a) $(a, b) \sim (a, b)$ (reflexivity)
(b) $(a, b) \sim (c, d)$ implies $(c, d) \sim (a, b)$ (symmetry)
(c) $(a, b) \sim (c, d)$ and $(c, d) \sim (e, f)$ implies that $(a, b) \sim (e, f)$ (transitivity)

Properties (a) and (b) seem so simple as to require no comment; (c) follows from 7 above. In fact, from $ad = bc$ and $cf = ed$ we obtain (by multiplying the first equation by e and the second by a)

$$ade = bce \text{ and } cfa = eda$$

and hence

$$bce = cfa$$

or

$$(be - fa)c = 0$$

Thus either $c = 0$, which in view of $b \neq 0$, $d \neq 0$ would imply that $a = 0$ and $e = 0$, or

$$be = af$$

Now consider a very general principle of wide applicability in mathematics.

Suppose we have a set S of objects $\alpha, \beta, \gamma, \ldots$ and that there is a relation R among these objects that is reflexive ($\alpha R \alpha$), symmetric ($\alpha R \beta$ implies $\beta R \alpha$), and transitive ($\alpha R \beta$ and $\beta R \gamma$ together imply that $\alpha R \gamma$). S splits into mutually disjoint classes such that objects in the same class are related through R while objects in different classes are not.

For example, let α, β, γ, ... be sets and $(\alpha R \beta)$ stand for the statement that there is a one-to-one correspondence between the elements of α and the elements of β. In this way a relation is defined between sets that is clearly reflexive, symmetric, and transitive. The collection of all sets now splits into disjoint classes, each class containing sets of the same cardinality.

Whitehead and Russell define cardinal numbers as classes of sets of the same cardinality. Thus number "three" is the class of all sets consisting of three elements, $\aleph_0$ (aleph null) the class of all denumerable sets, etc. Although it is a subtle bit of abstraction to identify numbers with certain equivalence classes, there are people who advocate introducing numbers in this way to children in the early grades.

Returning to our ordered pairs, we see that, since the relation $\sim$ of equivalence does have the three necessary properties, the set of all pairs also splits into disjoint "equivalence classes."

If we identify the ordered pair (a, b) $(b \neq 0)$ with the fraction a/b we see that the equivalence relation merely expresses the equality of fractions. Splitting the set of all ordered pairs into equivalence classes is simply assembling in each class all fractions that *represent* the same rational number.

But how can we say this before we have defined what we mean by a rational number? We cannot; but, since we really know what we are talking about, we can make everything perfectly legitimate by *identifying* rational numbers with the corresponding equivalence classes. In other words, *rational numbers are defined as classes of equivalent pairs.*

Let r and s be rational numbers; *i.e.*, suppose that r is a class of equivalent pairs and so is s.

To define $r + s$ we take a pair (a, b) in r (a so-called representative pair) and a pair (c, d) in s and construct the pair $(ad + bc, bd)$. (Note that $ad + bc$ is the numerator and bd the denominator of the fraction obtained by adding a/b and c/d according to the usual rules of addition.) Then $r + s$ is the class of pairs equivalent to $(ad + bc, bd)$. It may seem that this definition depends on the choice of representative pairs in r and s. However, it may be checked that if $(a'b')$ is also in r, *i.e.*, $(a', b') \sim (a, b)$ or $a'b = ab'$, and (c', d') in s, then $(a'd' + b'c', b'd')$ is equivalent to $(ad + bc, bd)$ and is hence in $r + s$. Similarly one defines rs as the class containing (ac, bd) if (a, b) is in r and (c, d) is in s.

What about our original integers? They now appear in a slightly disguised form as classes that contain pairs of the form $(a, 1)$. In other words, the integer a becomes the class of pairs equivalent to the pair $(a, 1)$.

While the equation $ax = b$ is not always solvable within the realm of integers (the equation $2x = 3$, for example, does not have an integral solution), it is easily solvable in terms of rational numbers. In fact, x is simply the class of pairs equivalent to (a, b). More than that, every equation $rx = s$, where $r \neq 0$ and where r and s are rational numbers, is now solvable in terms of rational numbers.

What we have done then has been to *embed* the integers in a larger set (*i.e.*, the set of rational numbers) in such a way that in this larger set the operation of division also becomes possible (the case of a zero divisor excluded, of course). In so doing we have preserved the integers and their operations while at the same time extending these operations, in a reasonable way, to all rational numbers.

In the preceding paragraphs we have described a general formal scheme for extending a system of objects on which operations of addition and multiplication can be performed to a system in which division also becomes possible. The objects need not be integers and the operations of addition and multiplication need not be familiar arithmetical operations. All that we require is that, *whatever the objects* and *whatever the operations*, conditions 1 to 7 (given above) be satisfied. As a matter of fact, one can dispense with some of them; in particular, condition 6 is not really needed. On the other hand, condition 7 is of crucial importance.

Let us now illustrate the great advantages of our formal approach by another example.

From now on our objects will be *continuous functions* $a(t)$, $b(t)$, . . . of the real variable t defined for $0 \leqslant t < \infty$. Addition is defined in the usual way but multiplication now will be the so-called convolution; *i.e.*,

$$a * b = \int_0^t a(t - \tau) \, b(\tau) \, d\tau = \int_0^t b(t - \tau) \, a(\tau) \, d\tau = b * a$$

For example, if $a(t) = 1$ and $b(t) = t$ then $a * b = \dfrac{t^2}{2}$ if $a(t) = t$ and $b(t) = \sin t$, $a * b = t - \sin t$, etc. It is worth noting that in general convolution with $a(t) \equiv 1$ is *equivalent* to integration from 0 to t. In this way the *transcendental* operation of integration takes on the *appearance* of an *algebraic* operation of multiplication. That there is more to it than appearance is shown briefly in the next few paragraphs. The operation of convolution appears frequently in many branches of pure and applied mathematics and therefore has been studied extensively. It has been realized for a long time that convolution resembles ordinary multiplication in many respects.

In fact it is easy for mathematicians to verify properties 1–5. Condition 6 does not hold (but as mentioned this is of no consequence); condition 7 can be proved, but it is by no means a simple matter. The standard way of expressing condition 7 is to say that there are no divisors of zero.

Now one can proceed exactly as before and extend the set of functions to the set of classes of equivalent pairs of functions and thus make the equation $a * x = b$ always solvable within this larger set.

The analogues of rational numbers now are certain operators that essentially are those introduced by Oliver Heaviside to solve linear differential equations encountered in the theory of electric circuits.

For example, the equation

$$\int_0^t x(\tau)d\tau = 1 \qquad t \geqslant 0$$

does not have a solution that is a continuous function of t. Formally though we can write

$$x = \frac{\{1\}}{\{1\}}$$

where $\{1\}$ denotes the function whose value for every $t \geqslant 0$ is 1. The division clearly is not ordinary division but an operation inverse to convolution. Now

$$x * f = \frac{\{1\}}{\{1\}} * f = \frac{\{1 * f\}}{\{1\}} = \frac{\int_0^t f(\tau)d\tau}{\{1\}} = f$$

so that x is simply the *operator* of *multiplication by the number 1*.

If we take the more general equation

$$1 * x = \int_0^t x(\tau)d\tau = a(t)$$

one can show that x is the operator s that is the sum of the differentiation operator D plus the operator $a(0) \frac{\{1\}}{\{1\}}$; $a(0)$ is the value of the function a at $t = 0$. Symbolically,

$$sa = Da + a(0) \frac{\{1\}}{\{1\}}$$

If p denotes the integration operator

$$pb = \int_0^t b(\tau)d\tau = 1 * b$$

one checks that

$$sp = ps = \text{identity operator; i.e.,}$$
$$(sp)a = (ps)a = a$$

for every function a that has a derivative.

In this way working with operators becomes completely analogous to manipulating ordinary fractions, but one must remember that multiplication is the operation of convolution.

The way we have introduced operators in analogy with the introduction of rational numbers is of relatively recent origin and is due to the Polish mathematician Jan Mikusinski. The use of operators to solve linear differential equations formally is much older and, as already mentioned, was initiated by Heaviside late in the 19th century.[3]

---

[3] Heaviside was criticized by some of his contemporaries for using formal manipulations without understanding how they worked. In reply to his critics Heaviside is quoted to have said "Should I refuse a good dinner simply because I do not understand the processes of digestion?" A similar criticism could be made of sixth-grade children learning the use of fractions without understanding the underlying theory.

We point this out because it tends to underscore one of the strong tendencies in contemporary mathematics: disregard and rejection of all that has not been logically formalized. This tendency (which is

Mikusinski's approach to Heaviside's calculus is an excellent example of what might be termed algebraization of mathematics. It is an outgrowth of a trend begun in the 19th century and continuing with strength and vigour to this day toward trying to fit mathematics into molds provided by abstract algebraic structures.

Algebraization has had its most spectacular success in topology. Next we give as an example of this Emil Artin's theory of braids.

Let L_1 and L_2 be two parallel straight lines in space identically oriented as indicated by the arrows (fig. 26).

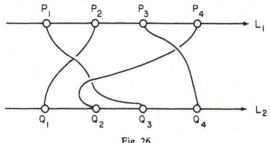

Fig. 26

On L_1 choose n distinct points $P_1, P_2, \ldots, P_n$ ordered in accordance with the arrow; similarly choose n points $Q_1, Q_2, \ldots, Q_n$ on L_2. From now on, for the sake of simplicity and definiteness, we take $n = 4$.

Each P is connected with a Q by a curve c that may wind and twist in space, but we shall require that its *projection* on the plane defined by the lines L_1 and L_2 be *monotone; i.e.,* as a point R moves along the projection from P to Q its distance from L_1 increases.

No two curves are allowed to intersect in space; in particular, no two curves can terminate in the same Q. In the figure we indicate that a curve is "above" or "below" another by an appropriate interruption of the projection of the one that is below.

What we have done so far has been to describe what may be called a *weaving pattern.*

To define a *braid* we now must introduce a class of *deformations* that, while capable of changing the *appearance* of the pattern, however drastically, leave its essential features unchanged.

---

beginning to permeate elementary and secondary school teaching) is, to an important éxtent, responsible for a growing separation of mathematics and physics. A physicist who uses mathematical descriptions is content to rely on internal agreement and, most importantly, on agreement with experiment. Like a sixth grader he is content to use rational numbers without full knowledge of how they can be incorporated into a formal system; like Heaviside he would happily juggle operators without waiting for a logical license to do so.

The deformations in question have the following properties:

(a) The lines L_1 and L_2 remain parallel with the same orientation though the distance between them can be increased or decreased at will.

(b) The points P and Q can be moved along their respective lines as long as their *order* is preserved.

(c) No two curves can intersect during deformation; they are "impenetrable."

(d) While the curves may be stretched or contracted at will, their projection on the (L_1, L_2) plane continues to have the monotonicity property stated above.

These properties are quite understandable if one imagines L_1 and L_2 as made of rigid material and the curves c as made of flexible and stretchable material.

We now call two patterns *equivalent* if one can be changed to the other by a deformation having the four properties stated above. This equivalence has the basic properties of reflexivity, symmetry, and transitivity; hence all weaving patterns fall into mutually exclusive classes of equivalent patterns.

A braid is such an equivalence class.

The fundamental problem of the theory of braids is to give a procedure (algorithm) that will make it possible to decide whether two braids are identical or not (or, what amounts to the same thing, whether the weaving patterns are equivalent or not).

This is a geometrical problem that properly belongs to the branch of geometry known as topology. In fact, topology can be defined as the study of properties of geometric configurations that remain unaltered (invariant) when the configuration is subjected to specific *continuous* deformations. Solution of the problem is accomplished by purely algebraic means that consist of defining a specific group and identifying each braid with an element of this group.

To indicate how this is done we first define the operation of *composition* of braids.

Let A and B be two braids (both with $n = 4$); to define $A \circ B$ we select a weaving pattern for A and a weaving pattern for B (fig. 27).

Let the lines, points, and curves for A be $(L_1, L_2, P_1, P_2, P_3, P_4, Q_1, Q_2, Q_3, Q_4, c_1, c_2, c_3, c_4)$; and let the lines, points and curves for B be $(L'_1, L'_2, P'_1, P'_2, P'_3, P'_4, Q'_1, Q'_2, Q'_3, Q'_4, c'_1, c'_2, c'_3, c'_4)$.

We deform B until the plane (L'_1, L'_2) becomes identical with the plane (L_1, L_2) and until L'_1 is made to coincide with L_2 (including orientation); care is taken to have L_1 and L'_2 on different sides of L_2.

We continue the deformation to make P'_1 coincide with Q_1, P'_2 with Q_2, etc. When this is done we remove (erase) L'_1 and hence L_2, "tying" the curves c and c' in an appropriate manner. The resulting pattern is a representative for the braid $A \circ B$.

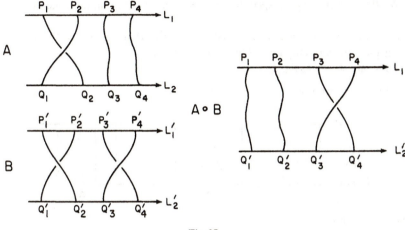

Fig. 27

A simple example of composition of braids is shown on fig. 27. The example is such that $A \circ B = B \circ A$; but this in general does *not* hold.

However, the operation of composition has the *associative property*

$$(A \circ B) \circ C = A \circ (B \circ C)$$

and there is a unique identity element I represented by fig. 28, the trivial braid

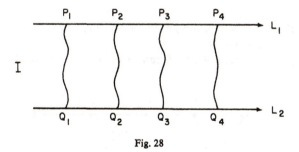

Fig. 28

One notes without difficulty that for every braid A one has

$$A \circ I = I \circ A = A$$

Every braid A has also a unique *inverse* A^{-1}; *i.e.*, the braid such that

$$A^{-1} \circ A = A \circ A^{-1} = I$$

To construct A^{-1} given A one simply exchanges "above" and "below"; *i.e.,* if a curve c_i is above c_j in A it is below c_j in A^{-1} and vice versa. Otherwise everything stays the same.

We thus see that with respect to the operation of composition braids form a group.

In this group there are three elements (braids) that together with their inverses play an especially important part in the theory.

These are shown as fig. 29.

(It may be seen that, *e.g.,* $A_1 \circ A_2 = A_2 \circ A_1$.)

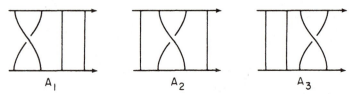

The inverses are:

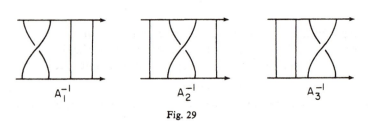

Fig. 29

The importance of these braids lies in the fact that *every braid can be composed of these basic elements.*

For example the pattern of fig. 26 can be deformed to look like:

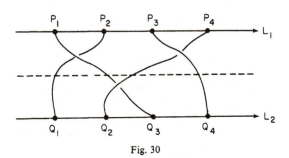

Fig. 30

A dotted line is drawn in fig. 30 to make it easier to see that the corresponding braid can be written as

$$A_1 \circ A_2^{-1} \circ A_3$$

Let us denote A_1^{-1}, A_2^{-1}, A_3^{-1} by B_1, B_2, B_3 respectively; with this notation every pattern is represented by a "word" like:

$$A_2 \circ A_1 \circ B_2 \circ B_3 \circ A_3 \circ A_1$$

Two patterns are equivalent if (and only if) the corresponding "words" represent the *same* group element.

For example, $A_1 \circ B_1$ and $A_2 \circ B_2$ are different "words" but they represent the same group element, namely the identity I.

In addition to the obvious relations

(a) $$A_1 \circ B_1 = A_2 \circ B_2 = A_3 \circ B_3 = I$$

the only other relation between the generating letters (or generators) is

(b) $$A_1 \circ A_3 = A_3 \circ A_1$$

The relation $B_1 \circ B_3 = B_3 \circ B_1$, for example, may seem to be another independent relation, but can be derived from (a) and (b) above.

One can now forget all about weaving patterns and braids and state the problem in purely algebraic terms as follows:

Given a group produced by six generators (A_1, A_2, A_3, B_1, B_2, B_3) that satisfy relations (a) and (b); is there an algorithm that will decide whether two "words" represent the same group element?

In the case of braids for arbitrary n such an algorithm is known and actually is relatively simple. But the closely related problem of classification of knots, though it leads to an analogous "word problem" for a similarly defined group, is up to now unsolved.

The theory of braids illustrates two important points:

The first is that problems that by their nature belong to the realm of the *continuous* can be successfully attacked and solved by methods that are inherently discrete and combinatorial. (Another instance of this is the use in Section 5a of Chapter 1 of the combinatorial lemma of Sperner in the proof of Brouwer's Fixed-point Theorem.)

The second point is that a problem may require invention of a new symbolism or formalism. If group theory were not known it is possible (though perhaps highly improbable) that it might have been invented for the purpose of solving the problem of classification of braids or a similar problem.

Mathematical creativity consists in either recognizing that an existing formalism is applicable to the problem at hand, or inventing a new one.

All that has been said above notwithstanding, if one had to name a single person whose work has had the most decisive influence on the present spirit of mathematics, it would almost surely be Georg Cantor. By the middle of the 19th century the accumulation of mathematical *material* had become so large and varied that the time was ripe for a synthesis and a reexamination of the foundations. Thus we witnessed the birth of the mathematical theory of sets on one hand and the study of mathematical systems (*i.e.,* mathematical logic) on the other.

In particular, mathematicians became more aware of the problem of *rigour* in the introduction of concepts and in the construction of proofs. Conscious examination of the foundations of analysis (*i.e.,* the infinitesimal calculus) and inquiry into the meaning of the real number system and of the nature of functions defined on them led to problems that are at the origin of modern set theory and of modern mathematical logic.

The great analysts and geometers following Newton (*e.g.,* Bernoulli, Euler, D'Alembert, Lagrange) had an almost unerring instinct in presenting valid theorems and proofs without a firm basis in formal systems and without strict adherence to standards of logical rigour. It hardly can be doubted that mathematical *intuition* (in the hands of people of genius) has such a clarity and unity that it anticipates special formalisms and makes them largely redundant.

The nature and origin of mathematical intuition pose philosophical and psychological problems. Perhaps in the distant future, when the nervous system and the organization of the human brain are better understood, some light may be thrown on such questions. If it turns out that the nature of logical thinking in mathematics is largely determined and influenced by this organization, then it perhaps may become clearer why intuitive mathematics is so naturally formalizable in an essentially unique way.

Like all great new theories, Cantor's set theory had precursors. Galileo already had remarked that the infinity of all integers is the same as the infinity of all the squares of integers because one can establish a one-to-one correspondence between the two sets. Giordano Bruno had considered explicitly an actual infinity of physical objects. But it was only in the middle of the 19th century that the mathematicians Bolzano, Weierstrass, and Dedekind and the logicians Boole, De Morgan, and, somewhat later, Frege and Peano, raised questions and constructed systems pointing toward the present edifice of set theory. However, the theory owes its full generality and scope to Cantor.

Cantor defined sets (*Mengen*) informally and did not employ any axiomatics in stating their properties. The fundamental idea he employed is that of one-to-one correspondence, and examples of sets were taken from the working mathematics of the period. Problems in the theory of Fourier series apparently led him to his general ideas. He observed that the set of all integers and the set of all rational numbers or of all algebraic numbers are of the same cardinality; *i.e.,* these sets can be brought into one-to-one correspondence with each other. Then came the all-important result: the power of the continuum of real numbers is greater than that of a countable set. This was established by the simple, but fundamental, diagonal method. There followed his papers on the theory of structure of point sets, and a new creation, the theory of ordinals. There then were beginnings of point set topology, real function theory, construction of the transfinite ordinals, a critical discussion of foundations of the infinitesimal calculus (with the rejection of the infinitesimally small), a critical and philosophical (rather than axiomatic) discussion of the nature of the continuum and of the foundations of mathematics in general.

Toward the end of Cantor's creative period came other important developments culminating in the formulation of the problem of the continuum. This celebrated problem of Cantor's can be formulated in a rough way as follows:

Is there a set which is more numerous than the set of integers but less numerous than the set of all real numbers?

The development of set theory really proceeded along two distinct lines although, very fortunately, they were strongly combined in Cantor's thinking.

One line pursued concepts of cardinality and order in an *abstract* fashion; *i.e.*, with little regard to the nature of sets. The other line concerned sets of points on a line, in a plane, or in higher-dimensional Euclidean spaces.

The first approach merged readily with logic; the second gave birth to point-set topology and was ultimately responsible for the fruitful theory of abstract spaces and for the all-important trend toward geometrization of mathematics.

Let us illustrate with an example the nature of what we call geometrization of mathematics.

Consider first numbers of the form

$$\frac{\epsilon_1}{2} + \frac{\epsilon_2}{2^2} + \frac{\epsilon_3}{2^3} + \cdots$$

where each ϵ can be either 0 or 1. To avoid duplication of certain numbers, *e.g.*,

$$\frac{1}{2} = \frac{1}{2} + \frac{0}{2^2} + \frac{0}{2^3} + \cdots = \frac{0}{2} + \frac{1}{2^2} + \frac{1}{2^3} + \cdots$$

we adopt a convention that in case of doubt we use the representation with infinitely many 0's (*i.e.*, the first of the two representations of ½ above). Then consider the set of numbers of the form

$$\frac{2\epsilon_1}{3} + \frac{2\epsilon_2}{3^2} + \frac{2\epsilon_3}{3^3} + \cdots$$

The two sets are in a clear one-to-one correspondence and are consequently of the same power; they even are of the same order type. The one-to-one correspondence preserves the order of points in both sets (*i.e.*, the relation of "greater than or equal to"). Abstractly they are quite indistinguishable, though their appearances as *sets of points* on a straight line are strikingly different.

The first set is the whole interval $(0, 1)$ including the endpoints. The second is the famous Cantor "middle-third" set that can be obtained by removing the open (*i.e.*, without endpoints) middle third of the $(0, 1)$ interval (*i.e.*, all points between ⅓ and ⅔) and then removing successively (ad infinitum) open middle-third portions of remaining intervals.

The Cantor set, though of the power of continuum, is very sparse. In fact, it is *nowhere dense* in the interval; *i.e.*, every point not in it can be enclosed in an interval that is also free from points of the Cantor set. The concept of a set *nowhere dense* in an interval contrasts with the concept of a set *everywhere dense* in the interval. A set is said to be *everywhere dense* in an interval if *every* subinterval contains at least one point (and hence infinitely many points!) of the set. Rational numbers between 0 and 1 form an everywhere-dense set in the $(0, 1)$ interval and yet, being denumerable, this set is less numerous than the Cantor set.

We thus see that one set can be numerous but sparse, while another can be scanty but dense. It should be clear that we are dealing with vastly different concepts. Cardinality and a power are extensions of the concept of *counting;* sparseness and density have something to do with *spatial arrangements* and "proximity."

By combining the two concepts one arrives at a very satisfactory definition of "small" sets.

A set is said to be of *first category* (or "small") if it is a union of *denumerably many* (or fewer) *nowhere-dense sets*. In this way both the set of rational numbers and the Cantor middle-third set are "small"; the first because it is scanty and the second because it is sparse.

But now one goes much further.

First one analyzes what is involved in the concept of denseness and one concludes easily that what one really needs is the concept of a *neighbourhood of a point*, generalizing the concept of an interval. This concept in turn can be based on the concept of distance. (An interval with midpoint x_0 can be defined as the set of points x whose distance from x_0 does not exceed a prescribed positive number δ; this definition in higher-dimensional spaces yields spheres as analogues of intervals.)

It turns out that for many purposes one needs only three properties of the distance:

(a) The distance between a and b is nonnegative and zero only if a and b are identical.

(b) The distance between a and b is the same as between b and a.

(c) The distance between a and b is less than or equal to the sum of the distances from a to c and from b to c whatever c is. (This is the so-called triangle inequality.)

Consider, for instance, the set C of all continuous functions of the real variable t defined for $0 \leqslant t \leqslant 1$.

Given two such functions $a(t)$ and $b(t)$ one defines the "distance" $\delta(a,b)$ between them by the formula

$$\delta(a, b) = \max_{0 \leqslant t \leqslant 1} \left| a(t) - b(t) \right|$$

and one verifies that the basic conditions (a), (b), and (c) above are satisfied.

Many purely analytic facts about continuous functions now can be couched in geometric terms. For example, the famous theorem of Weierstrass that every continuous function can be approximated uniformly, with arbitrary accuracy, by polynomials can be restated by saying that the set of polynomials is everywhere dense in the space C of continuous functions.

Such seemingly purely verbal restatements have proved enormously stimulating as sources of new (geometric) insights and of new problems.

As an example we quote the remarkable result of Banach that the set of continuous functions possessing a derivative at *at least one point* forms a set of first category in the space C of all continuous functions. In other words, continuous nowhere-differentiable functions are not just pathological creatures but constitute an *overwhelming majority* of all continuous functions! (This is because they form a set of second category, *i.e.*, a complement of a "sparse" set of first category.)

It is perhaps even more remarkable that it is almost easier to prove that "most" continuous functions are nowhere differentiable than to exhibit an explicit example of one such function!

Thus the concept of sets of first category became a powerful tool in proving the existence of certain mathematical objects.

Apart from Cantor's monumental work, there were other motives for the development of axiomatic method. After the discovery and development of non-Euclidean geometries, impetus was given to the establishment of axiomatic systems of geometries that are more general than Euclid's and that also embrace more qualitative geometric systems like projective geometry. The work of Hilbert mentioned above and in particular that of the American geometers like Veblen, brought new interest and scope to the employment of axiomatic thinking in other parts of mathematics. Peano's work on axiomatization of *arithmetic,* together with that of Boole on algebra of sets, was an essentially parallel development. After the creation of set theory, it became appropriate, even imperative, to attempt the construction of a system of axioms for the whole of mathematics. The elegance and success of the axiomatic method in individual parts of mathematics encouraged these attempts. In fact, works such as that of Whitehead and Russell were greatly influenced by the experience gained in dealing with the axiomatic systems in parts of geometry, arithmetic, and algebra.

The great program of Hilbert was to erect an axiomatic edifice, sufficient for all work in mathematics. Not that there was unanimity as to the admissibility or sense of all axioms! In particular, the axiom of choice was felt by some mathematicians to be of dubious character and perhaps inadmissible because of the strange, seemingly paradoxical consequences of its application. (We already have mentioned the paradoxical decompositions of spheres of different radii into a finite number of mutually congruent subsets.) A debate continued at the beginning of the 20th century as to the role and meaning of this axiom. It must be said that throughout the history of mathematics new objects constantly were being discovered with properties to which the mathematical thinking of the period was unaccustomed—even without the use of axioms (like that of choice) stating the existence of "nonconstructive" entities. The process of generalization in mathematics very often has started from such "surprising" discoveries. Their logical consequences, no matter how strange they might have appeared at the moment, had to be accepted and often have formed a basis for new systems. The school of intuitionists, headed by L. E. J. Brouwer and for a time by Lebesgue and H. Weyl, have attempted to confine mathematics to more constructive or "operational" systems. The great majority of mathematicians, however, did not reject the axiom of choice.

Hilbert's program implied a faith in the completeness of an all-embracing axiomatic system for the whole of mathematics. The work of Bernays, Fraenkel, and von Neumann already had laid solid foundations for axiomatic systems of set theory and mathematical logic. There was reason to hope that all meaningful problems in such systems were (in principle) decidable.

Then in 1931 Gödel published his paper, *Über formal unentscheidbare Sätze der Principia Mathematica and verwandter Systeme I* ("On Formally Undecidable Statements of the *Principia Mathematica* and Related Systems I"). Speaking broadly, his result is that in any sufficiently rich system of axioms (in fact, rich enough to include arithmetic) there will exist statements that, though meaningful, are undecidable within the system. Even more surprisingly, these statements can be shown to hold or be "true" in the sense that they appear in the form of assertions that all integers possess certain arithmetic properties and these can be verified for every integer that is examined.

That "true" statements may not be provable can be illustrated with the following example:

Consider the statement that *for every* positive integer n one has

$$1 + 2 + \cdots + n = \frac{n(n + 1)}{2}$$

This is usually proved by applying the axiom of mathematical induction that states: if a statement $S(n)$ concerning positive integers is true for $n = 1$ and if the truth of $S(m)$ implies the truth of $S(m + 1)$ then $S(n)$ is true for *all* integers n. This axiom *allows* us then to make a statement about an infinity of objects (positive integers in our case) without performing infinitely many verifications; clearly an impossible task. If the axiom of induction were not available the statement that *for every n* one has

$$1 + 2 + \cdots + n = \frac{n(n + 1)}{2}$$

could not be proved because, of all the axioms of arithmetic and logic, the axiom of induction is the *only* licence to deal with the whole infinite set of integers.

One might try to escape the yoke of induction by attempting an indirect proof. One could say that if $1 + 2 + \cdots + n$ were not equal to $n(n + 1)/2$ for some n then there would be the smallest such n; this could not be equal to 1 because our statement is "true" for $n = 1$. It could not be greater than 1 because one could then show that $n = 1$ is also an exception in contradiction to n being the smallest one. Alas, the reasoning is based on the principle that every nonempty set of integers has a smallest element and this happens to be *equivalent* to the axiom of induction.

Without the axiom of induction then there would be simple arithmetic statements like

$$1 + 2 + \cdots + n = \frac{n(n + 1)}{2}$$

which even though "true" could not be derived from the remaining axioms; so that one could say that without the axiom of induction arithmetic is *incomplete*.

What Gödel has shown is that *every* sufficiently rich system of axioms is incomplete and that it cannot be made complete by the addition of any finite number of new axioms.

It is impossible to give here a detailed account of Gödel's proof, but we will give a sketchy account. He began by an enumeration of all allowed mathematical statements that employ the prescribed symbols and rules of operation of the system; this class is countable and Gödel attached an integer to each such expression. Every mathematical proposition is thus reduced to a statement about integers. By a process similar to the diagonal construction of Cantor (mentioned earlier), Gödel exhibited a statement within the system that cannot be proved or disproved with the means of the system itself. (The theorem is applicable, of course, only to systems that are *consistent; i.e.,* within which a contradictory statement like $1 = 0$ cannot be proved.)

To explain a little more fully the ideas that underlie Gödel's construction we must first discuss the concept of a *formal system*.

A formal system consists of a finite set of symbols and of a finite number of rules by which these symbols can be combined into formulas or statements. A number of such statements are designated as axioms and by repeated applications of the rules of the system one obtains an ever growing body of demonstrable (provable) statements.

A proof of a given statement (a formula) is a finite sequence of statements that starts with an axiom and ends with the desired statement. The sequence is such that every intermediate statement is either an axiom or is derivable by the rules of the system from statements that precede it.

However, a statement that a sequence of formulas does or does not constitute a proof of a formula is *not* a statement in the formal system itself. It is a statement *about* the system and such statements are often referred to as *metamathematical*.

Failure to distinguish carefully between *mathematical* and *metamathematical* statements leads to paradoxes; the earliest of these is the "All Cretans lie" paradox of Epimenides the Cretan.

Much more decisive for our purposes is the following variant of a famous paradox of Jules Richard.

We say that a function f defined for all integers $n = 0, 1, 2, \ldots$, and whose values also are nonnegative integers, is *computable* if there is a prescription containing a finite number of words that then allows computing $f(n)$ (the value of f for the integer n) in a finite number of steps. This may, and indeed almost always does, depend on n.

The set of all computable functions easily can be seen to be countable (denumerable). Hence it is possible to arrange all computable functions in a sequence $f_1, f_2, f_3, \ldots$.

Define now a new function g by the formula

$$g(n) = f_n(n) + 1$$

This function is not contained in the above sequence because for $n = 1$ it differs from $f_1(1)$, for $n = 2$ it differs from $f_2(2)$, etc. Hence it is *not* computable.

On the other hand it is clearly computable, for $f_n(n)$ is computable and by adding 1 to it one obtains $g(n)$.

The origin of this paradox is clear enough. The construction of g depends in an essential way on the ordering of the f's; although the f's are described within a system (*e.g.*, arithmetic) their ordering is a *metamathematical operation*.

Imagine now that somehow the metamathematical statement "m is the value which the nth function in the sequence $f_1, f_2, \ldots$ assumes at n" could be translated into a purely arithmetic statement; *i.e.*, a legitimate statement within the system. Then, assuming that the system is free from contradiction, the computability of g would be *undecidable*.

Gödel's great idea was to translate metamathematical statements into statements of arithmetic; thus *mirroring* them, as it were, within the formal system. Since mathematical and metamathematical statements now could be combined freely within the system, questions that in the ordinary course of events would lead to a paradox would be mirrored into *undecidable propositions*.

A few words on the Gödel numbering are in order.

A formula in a formal system that includes arithmetic contains such fixed signs as $\supset$ implication sign, 0 zero, $\vee$ logical "or," and (left parenthesis, as well as *numerical* variables $x,y,z, \ldots$ for which nonnegative integers can be substituted, propositional (sentential) variables $p,q,r, \ldots$ for which propositions (sentences) can be substituted, and *predicate* variables $P,Q,R \ldots$ for which such predicates as "composite" or "greater than" can be substituted.

One can get along with ten fixed signs; these are assigned numbers from 1 to 10 (*e.g.*, $\supset$ is assigned number 3, 0 number 6, and so on).

Numerical variables are assigned primes greater than 10 (*e.g.*, x is assigned 11, y 13, and so on), propositional variables are assigned squares of primes greater than 10 (*e.g.*, p is 11^2, q 13^2, etc.), and predicate variables are assigned cubes of primes greater than 10.

A formula is then assigned a number according to a rule that is best described in an illustrative example. Consider the formula

$$(x > y) \supset (x = sy) \vee (x > sy)$$

which states that x greater than y implies that either x is an immediate successor of y (*i.e.*, $x = y + 1$) or x is greater than the immediate successor of y. The formula contains nineteen symbols: $(,x, >, y,), \supset, (,x, =, , s,y,), \vee, (,x, >, s,y,)$, some (like x or s) with repetitions. We then take the first nineteen primes, raise each of them to the power assigned to the corresponding symbol, and multiply: The resulting number

$$2^8 \times 3^{11} \times 5^{13^3} \times 7^{13} \times 11^9 \times 13^3 \times 17^8 \times 19^{13} \times 23^6 \times 29^7 \times$$
$$31^{13} \times 37^9 \times 41^2 \times 43^8 \times 47^{11} \times 53^{13^3} \times 59^7 \times 61^{13} \times 67^9$$

is the Gödel number of our formula. The reader should note that the left parenthesis (is assigned number 8, the right parenthesis) 9, $\vee$ number 2, s number 7, and the predicate $>$ the number 13^3.

A sequence of formulas (such as may constitute a proof) $F_1,F_2,F_3, \ldots, F_m$ is assigned the Gödel number

$$2^{G_1} \times 3^{G_2} \times \cdots p_m^{G_m}$$

where p_m is the mth prime and $G_1,G_2, \ldots$ are the Gödel numbers of the formulas $F_1,F_2, \ldots$ respectively.

In this way a unique integer corresponds to each formula or sequence of formulas. Although not every integer is a Gödel number, if it is, it determines the expression *uniquely*. This follows from the unique factorization theorem: except for order, there is only one way in which an integer greater than 1 can be written as a product of powers of primes.

Note also that a metamathematical statement "$(x > y)$ is an initial part of the formula $(x > y) \supset (x = sy) \lor (x > sy)$" is mirrored within the system in the purely arithmetical statement that the Gödel number of $(x > y)$ which is

$$2^8 \times 3^{11} \times 5^{11^3} \times 7^{13} \times 11^9$$

is a *divisor* of the Gödel number of the full formula.

In general, Gödel's constructions lead to statements *within* the system that allow an interpretation *about* the system. As mentioned above such statements in the absence of proper precautions may lead to paradoxes. Gödel shows how *with* proper precautions these paradoxes turn into *undecidable propositions*.

That Gödel's discovery should have produced a revolutionary change in mathematical logic is clear. But it went far beyond that by producing a profound change in the philosophical outlook of the whole of mathematics.

To appreciate this one must realize that the undecidable propositions of Gödel were not some esoteric statements far removed from the mainstream of mathematics but that, owing to the idea of numbering, they could be stated in terms of *Diophantine equations* that for centuries have been *bona fide* objects of purely mathematical investigations.

Diophantine equations are ordinary algebraic equations in one or more unknowns. The problem is: do they or do they not have solutions that are integers?

Many famous unsolved problems of mathematics refer to Diophantine equations. Of these perhaps the most celebrated is Fermat's problem to prove that for $n > 2$ the equation

$$x^n + y^n = z^n$$

has no solutions in integers. For $n = 2$ there are infinitely many solutions: such so-called Pythagorean numbers as $(3,4,5)$, $(12,5,13)$, etc. For $3 \leqslant n \leqslant 100$ Kummer proved Fermat's conjecture, and there are many closely related results.

Could it be that this problem is undecidable in the present system of mathematics?

Such questions would never have arisen in mathematics before Gödel's monumental discovery. Because of Gödel, logic was lifted from its accustomed place at the foundations and injected into many of the problems and preoccupations of everyday mathematics.

Speaking somewhat generally, there are two distinct kinds of mathematical arguments: (a) existential and (b) constructive. In Chapter 1 we have seen some examples of both. For instance, one can use Cantor's argument to prove that there exist transcendental numbers without exhibiting a single example, or one can use Liouville's construction to produce a whole class of *concrete* transcendental numbers.

Similarly, one uses a *reductio ad absurdum* argument to prove that every algebraic equation of degree n with complex coefficients has at least one complex root; such a proof is of little use, though, if one happens to be interested in the numerical value of the root.

The possibility of proving the existence of objects without being called upon to *exhibit* them is one of the most distinctive features of mathematics. But purely existential arguments can be carried to the point that they cause a feeling of uneasiness.

We have already mentioned that some quite innocent-sounding axioms like the axiom of choice (given a collection of nonoverlapping nonempty sets one can form a new set by choosing one element from each set of the collection) allow one to prove the existence of objects (*e.g.*, nonmeasurable sets) that are so strange that they defy all intuition.

Perhaps the only philosophical issue that causes a serious division among mathematicians concerns their attitude toward the *existence* of mathematical objects. There is universal agreement, however, that *algorithms* or *constructive procedures* are of great importance and interest.

One can define the most general algorithm in terms of so-called recursive functions. Rather than do this, we shall describe one nontrivial special algorithm in the hope that it exhibits the essential features of all algorithms.

Our example is Euclid's algorithm for finding solutions in integers of the Diophantine equation

$$ax + by = 1$$

where a and b are nonnegative integers.

First we note that if a and b have a divisor $d > 1$ in common, then the equation has no solutions. (If there were, the left-hand side would be divisible by d; hence 1 would be divisible by $d > 1$ which is clearly impossible.) We thus may assume that a and b are relatively prime; *i.e.*, they have no divisor in common except 1.

Next we note that if either a or b is 1 we immediately have a solution: $x = 1$, $y = 0$ ($a = 1$) or $x = 0, y = 1$ ($b = 1$).

If a and b are relatively prime and neither is equal to 1 (this implies, in particular, that $a \neq b$) Euclid's algorithm consists of the following set of directions:

(1) Suppose that a is larger than b and divide a by b; this gives the quotient q_1 and the remainder r_1 which is *less than* b. In other words,

$$a = q_1 b + r_1 \qquad 1 \leqslant r_1 < b$$

Substitute this into the equation obtaining

$$r_1 x + b(q_1 x + y) = 1$$

so that setting $x = x_1$ and $q_1 x + y = y_1$ we have

$$r_1 x_1 + b y_1 = 1$$

(2) Note that the new equation is of the same form as the original except that r_1 is less than a (since it is less than b which is less than a).

Divide b by r_1; *i.e.*, write

$$b = q_2 r_1 + r_2 \qquad 1 \leqslant r_2 < r_1$$

and substitute this into the new equation obtaining

$$r_1(x_1 + q_2 y_1) + r_2 y_1 = 1$$

Setting

$$x_2 = x_1 + q_2 y_1, \ y_2 = y_1$$

we get

$$r_1 x_2 + r_2 y_2 = 1$$

which is again of the original form.

(3) Continue the process until either the coefficient of some x_k or that of some y_k becomes 1.

Take the $(1,0)$ or $(0,1)$ solution of this equation and by retracing the steps find a solution to the original equation.

Here is a simple numerical example

$$14x + 9y = 1$$
$$14 = 9 + 5$$
$$5x + 9(x + y) = 1$$
$$5x_1 + 9y_1 = 1 \qquad\qquad x_1 = x, \ y_1 = x + y$$
$$9 = 5 + 4$$
$$5(x_1 + y_1) + 4y_1 = 1$$
$$5x_2 + 4y_2 = 1 \qquad\qquad x_2 = x_1 + y_1, \ y_2 = y_1$$
$$5 = 4 + 1$$
$$x_2 + 4(x_2 + y_2) = 1$$
$$x_3 + 4y_3 = 1 \qquad\qquad x_3 = x_2, \ y_3 = x_2 + y_2$$
$$x_3 = 1, \ y_3 = 0$$
$$x_2 = 1, \ y_2 = -1$$
$$x_1 = 2, \ y_1 = -1$$
$$x = 2, \ y = -3$$

Similar algorithms can be devised to produce solutions of other Diophantine equations; *e.g.*,

$$x^2 - 2y^2 = 1$$

But it is an open question if there is an algorithm for determining whether there exist solutions of Diophantine equations of arbitrarily high orders in an arbitrarily high number of variables; *i.e.*, equations of the form:

$$\sum_{i_1 i_2, \ldots, \ i_k = 0}^{n} a_{i_1 i_2} \ldots {}_{i_k} x_1^{i_1} x_2^{i_2} \ldots x_k^{i_k} = 0$$

where the a's are integers. In 1900 Hilbert presented to the International Congress of Mathematicians in Paris a list of problems that since has become justly famous. Some have been solved in the meantime; the problem in question is Hilbert's tenth problem.

We can see how close this problem is to that of decidability and related logical problems; it should therefore not come as a surprise to learn that much progress has been made in recent years toward solving this problem by methods inspired by mathematical logic.

An algorithm is a set of precise instructions telling how to perform a certain task. It is easy to conceive an automaton that will perform according to an algorithm without human intervention. Is there, however, a universal automaton that could be programmed to execute *any* algorithm?

An affirmative answer to the question was given by the English mathematician Alan Turing, and his work became the theoretical basis for modern, all-purpose digital computers.

Turing's universal machine is actually quite simple. It consists (fig. 31) of an infinite tape divided into equal squares and

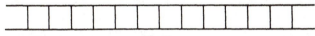

Fig. 31

a finite set of symbols (alphabet) which for the sake of simplicity may be taken as consisting of only one symbol: a vertical dash |.[4]

There is also a movable *scanning square* that allows one to scan the tape square by square. Finally the machine can perform the following operations:

- (l): move scanning square one unit to the left
- (r): move scanning square one unit to the right
- (R): replace the symbol in the scanning square by any other symbol of the alphabet (in our case this means either erasing a vertical dash or printing a vertical dash in a blank square)
- (h): halt the procedure

A program is a set of instructions of the following form: "if the symbol in the scanned square is _____ perform _____ and look up instruction _____."

As an example, here is a program for finding the remainder after dividing an integer by 3. An integer n is represented on the tape as n vertical dashes placed in successive squares, and the scanning square is placed to the right of the dashes. The numbers refer to instructions and * stands for "blank." For ex-

---

[4] If one wants to be fussy one should say that there are actually two symbols, the other symbol being the blank.

ample, an instruction may read as follows: "if the scanned square is blank halt the operation but if it is not, move to the left and look up instruction 2."

```
0 * 1 0
0 | 1 1
1 * h
1 | 1 2
2 * h
2 | R 3
3 * r 4
4 | R 5
5 * r 6
6 | R 0
```

It is actually possible (indeed it is done in real computers) to store the instructions on the tape in an appropriately coded form; but, had we done this, the program would become much more complicated without adding a great deal to the understanding of the principles involved.

We see that a Turing machine is a very simple formal system that is nevertheless rich enough to be able to reproduce all possible algorithms. As a counterpart of Gödel's result one can show that there are simple arithmetic and combinatorial questions that are *algorithmically undecidable; i.e.,* there is no program of a Turing machine that will decide the truth or falsity of these questions.

An example of such an undecidable question is the general "word problem" in group theory.

Suppose that we start with four abstract symbols A, B, A′, B′, that can be juxtaposed to form arbitrarily long "words"; *e.g.,*

$$ABB'AAA'BBA'B$$

Suppose also that a finite number of such words are assumed to be equal to the "identity element" *I; i.e.,* whenever such words are encountered as parts of other words they can be removed, thus shortening the original words.

If, for instance, we assume that

$$AA' = A'A = BB' = B'B = I$$

and also that

$$B'AA = I$$

the word ABB′AAA′BBA′B can be shortened either to ABB′ABBA′B (using AA′ = I) or to ABA′BBA′B (using B′AA = I).

Given such a scheme, the problem is: Does there exist an algorithm that will decide if two words are or are not identical (*i.e.,* if they represent the same group element)? In *general* the answer is no, as was shown by the Soviet mathematician Novikov. On the other hand there are many special groups (as defined by the finite number of relations between its generators) for which the word problem is decidable.

A trivial example is the group defined by the relations

$$AA' = A'A = BB' = B'B = I$$

and $$ABA'B' = I$$

(This is a disguise for AB = BA, so that the group is commutative.)

Here the criterion for deciding whether two words are identical is especially simple. One counts the excess of A over A' characters and the excess of B over B' characters in both words; if the two excesses are the same for both words the words are identical and vice versa.

A more difficult example is provided by the groups associated with braids (discussed earlier in this chapter); the word problem for these groups is decidable and therefore the decision whether two braids are identical can be left to a computing machine (*e.g.*, a Turing machine).

The subject of computing machines and their role in mathematics is still under debate. Mathematicians exhibit a gamut of attitudes ranging from indifference to hostility; a small number feel that computing machines are destined to play a significant role in the future development of mathematics above and beyond their unchallenged usefulness and power as tools of science and technology.

The idea of using mechanical devices to perform arithmetical operations and aid in lengthy computations is very old. The ancients had constructed simple devices and thought of graphical methods that would save time by performing simple mathematical steps faster than they could be written by hand. We cannot resist quoting here from Plutarch's Life of Marcellus: "Eudoxus and Archytas had been the first originators of this far-famed and highly prized art of mechanics, which they employed as an elegant illustration of geometrical truths, and as a means of sustaining experimentally, to the satisfaction of the senses, conclusions too intricate for proof by words and diagrams. In the solution of the problem, so often required in constructing geometrical figures, given the two extremes, to find the two mean lines of a proportion, both these mathematicians had recourse to the aid of an instrument, adapting to their purpose certain curves and sections of lines. But what of Plato's indignation at it, and his invectives against it as mere corruption and annihilation of the one good in geometry, which was thus shamefully turning its back upon the unembodied objects of pure intelligence to recur to sensation, and to ask help (not to be obtained without base supervisions and depravation) from matter; so it was that mechanics came to be separated from geometry, and, repudiated and neglected by philosophers, took its place as a military art."

Pascal constructed an arithmetical machine; Leibniz envisaged a logical machine that ultimately could treat all the problems of the mathematical sciences.

By conceiving both the idea of a formal system and the possibility of operating in it mechanically, Leibniz had in mind really something very close to what has developed in recent times. The 19th century saw the development of mechanical devices for computing through the work of Babbage and others; but it is only in very recent times (in fact, after about 1940) that electronic technology has been able to provide the means of starting a development on a scale never contemplated before. It promises greatly to enlarge the scope of mathematical material and ultimately to influence the directions and pace of mathematical research itself.

Throughout the history of mathematics new ideas have come both from flashes of intuition and from patient observation. Perusal and discussion of the accumulating facts suggested new generalizations. In number theory especially, properties of integers were often first observed by experimentation with "small" numbers. Gauss himself, when asked how he divined some of his general ideas, replied: *"Durch planmässiges Tattonieren"*—through systematic, palpable experimentation. It is hard to exaggerate the suggestive role of examples and hints contained in special cases in determining the *directions* that mathematicians take in their research.

Machines available in the late 1960s perform the arithmetical operations of addition, subtraction, multiplication, and division on two numbers, to precisions of 1 part in 10^{12}, in less than one-millionth of a second. They are provided with *memories* (that is, storage devices) in which hundreds of thousands of such numbers can be stored and rapidly retrieved. The time involved in entering or retrieving data from the memory is also only of the order of one-millionth of a second. The devices are able to perform simple logical processes (Boolean operations) and are provided with many "orders" that automatically execute simple combinatorial steps.

Design and construction of machines as well as development of methods for presenting mathematical problems efficiently and accurately was stimulated by the technological problems of World War II. Work on problems in pure science has been steadily increasing in scope and in substance. So far, most of this work concerns mathematical problems of theoretical physics. In mathematics itself, it chiefly concerns combinatorial analysis and number-theoretical questions.

It might appear that really interesting questions require dealing with numbers so large that the limitations of machine memory would prohibit their useful study. However, in many cases, formulas expressing limiting or asymptotic behaviour of functions already are well illustrated for small or moderate values of their arguments. For example, the density of primes up to 10,000 gives a fair picture of their asymptotic density. We already have mentioned the Prime-number Theorem. This states that the number $\pi(n)$ of primes from 1 to n is asymptotically equal to $n/\log n$. This means that, roughly speaking, for large n there will be about one prime number between n and $(n + [\log n])$ where $[\log n]$ denotes the largest integer smaller than $\log n$. Sometimes there is exactly one prime in this range, sometimes there will be two primes within it, sometimes three, and sometimes none. It is reasonable to divide integers into classes, with C_0 the class of all integers n for which there is no prime in the range defined above, C_1 the class of all integers for which there is exactly one prime in the range, C_2 the class of those for which there are two primes, and so on. It seems beyond the present means of the theory of numbers to show that these classes possess asymptotic densities; *e.g.*, to prove that if we take the number $\gamma_0(N)$ of integers from 1 to N, which belongs to class C_0, then the limit $\gamma_0 = \lim\limits_{N \to \infty} \dfrac{\gamma_0(N)}{N}$ exists. With a modern computing machine that can store all the primes up to 100,000,000, it is an easy matter to investigate the behaviour of these ratios. It seems that they approach definite limits, the first of which (γ_0) appears to be about .30 . . . ; the others come out to be $\gamma_1 \cong .42 . . . ; \gamma_2 \cong .21$ . . . ; $\gamma_3 \cong .05 . . . , \gamma_4 \cong .006 . . .$, etc. Here a simple observation suggests theorems yet unproved; first that the limits exist and secondly that these limits tend to 0 with increasing index of the class.

In combinatorial analysis (*i.e.*, the study of patterns, and their growth and development) such heuristic work already has proved to be of considerable value. For example, a conjecture of Euler on mutually orthogonal Latin squares was recently disproved by an example suggested by numerical work on computers.

Questions of enumeration in combinatorial analysis may be studied by "brute force" on a computer. As in the number-theoretic example above, the computer can examine all possible cases for small values of the variable n and by inspecting the dependence on n may aid in formulating conjectures about asymptotic behaviour. As an example, we may mention the following: Given the set E of integers from 1 to n and two permutations S_1 and S_2 of this set, which are selected *at random;* what is the expected number of elements in the group generated by these permutations? To approach such problems through work on an electronic computer, one might use a statistical sampling procedure known as the Monte Carlo method. The idea behind this approach is extremely simple; its usefulness would be nil were it not for high-speed computers.

Here is how it works. Let $g(S_1, S_2)$ denote the number of elements in the group generated by S_1 and S_2 (*i.e.*, the number of *different* permutations obtained by forming all possible products like $S_1^{k_1}, S_1^{l_1}, S_1^{k_2}, S_1^{l_2} . . .$). The desired average is simply

$$\frac{\Sigma g(S_1, S_2)}{\text{number of distinct pairs } (S_1, S_2)}$$

where in the numerator Σ denotes summation over all distinct pairs S_1, S_2.

The denominator is seen to be

$$\frac{(n!)^2 + n!}{2}$$

which for n = 5 is already sizable—7,260.

Thus even for *n* so small, one must examine 7,260 groups and find their orders. The task is not unlike that of trying to determine the average weight of newborn male children. Rather than average the weights of *all* newborn male children one picks a (representative) *random sample*. How to do this without introducing all sorts of biases is part of the art and science of sampling. A computer can be instructed to draw random samples from the set of all pairs of permutations. One gets an excellent approximation to the desired average by this technique.

But wider vistas have been opened by the development of fast electronic machines. Beyond the general possibilities of large-scale experimentation in problems of pure mathematics or in the exploration of tentative ideas in physical theories, it is possible to envisage machines actually doing work in formal systems of mathematics. Speaking broadly, the present machines operate on a set of given instructions (a flow diagram and a code) that, once given, make the machine proceed automatically in solving numerical or combinatorial problems. The course of the operation is completely prescribed; the only flexibility left to the machine consists in choosing one course of the calculation, or another, depending on the values of numbers just computed. Until now, the so-called decisions made by the machine involve in practice only a limited set of changes in the logical course of computation.

It is possible to conceive of a more general plan: If a machine should be kept in constant communication with an intelligent operator, who could change the logical nature of the problem itself during the course of computation (depending on his interpretation of results and on his observations), a much greater scope of exploration would be opened. If fast communication between the operator and the machine were possible, not only could the drudgery of elementary algebraic or analytic calculations be taken over by the computer but also, for example, in the search for examples or counterexamples, the machine could quickly provide and display visually on a screen the elements envisaged by the working mathematician and guide or verify his intuition.

As an example, we may mention the study of properties of functions of several variables and of transformations of the space of several variables into itself.

In many problems it is important to find the critical values of a function of several real variables $f(x_1, x_2, \ldots, x_n)$, the function f being given analytically; *e.g.*, in terms of elementary functions. It is well known that the usual procedure employed to find the actual numerical values of local minima or maxima of a single function requires a search that is extremely time-consuming. If the number of independent variables is large, for example 5 or more, no really efficient method of finding all the critical points is available. Imagine now that the values of the function can be quickly computed on a grid of points that forms, say, a two-dimensional section in the given space. The resulting "graph" of the function of two variables (*i.e.*, a surface) is projected with a cathode-ray display tube. (A code for calculating axonometric projections is readily made.) One quickly sees the region or regions where minima are likely to be assumed. By a quick change of scale and a magnification (that is, a subdivision of the region in question into a greater number of points), a computing machine can act as a microscope of arbitrary power. All we are saying is that, instead of the blind recipes embodied in a search code for critical points, one can utilize human visual perception, which is still much quicker than any known automatic code for "recognition."

For functions of 3 or 4 variables one also should have a quick way to instruct a machine to select a desired two-dimensional section, to establish on it a set of independent grid points, to compute the value of functions on these points, and to display it in perspective on a screen. Equally important will be the ability to change the scale of the independent and dependent variables by a general linear transformation.

The next aim, still more ambitious, would be to provide for a series of "experiences" with problems computed on the machine so that the operator would acquire a *feeling*, after some practice, for the four-dimensional space as a result of such experimentation. Let us consider the problem, in three dimensions, of threading a solid through a closed space curve, an exercise that involves trial and error. No simple criteria concerning projections seem to be sufficient to decide whether or not one can push a given solid through a given curve. The physical process of effecting such penetrations could be imitated by the machine by making it compute successive positions of the solid following given manual instructions (to be quickly transmitted numerically) about rotations and translations in three-dimensional space. The contact between the two sets would be tested after each trial displacement.

Programs enabling the machines to work with symbolic expressions, devising ways to deal with certain simple systems of axioms in producing formal proofs and searching for theorems, already have been set up. This work is only at its beginning. Clearly, one can operate formally on polynomials and rational fractions by coding ways to perform algebraic operations on them. There are programs to make machines perform formal differentiation and search for indefinite integrals of a large class of elementary functions. There also exist interesting programs for proving theorems in systems of Euclidean geometry or in projective geometry. Machines already have produced amusing proofs of properties of triangles, etc., sometimes different from the familiar ones of school experience.

In their work on logic, so far the machines have been limited to elementary Boolean operations or, equivalently, to the sentential calculus. The next step would be to make them operate with quantifiers "there exists an x such that ..." and "for all x . . . . . ," thus vastly enlarging their logical capability.

Of real interest is the fact that computing machines have proved to be a source of new, interesting, and challenging mathematical problems.

As an example we mention a class of problems dealing with so-called *algorithmic complexity*.

It somehow seems that multiplication is a more complex operation than addition. Is there a way of translating this vague feeling into a precise mathematical statement?

A reasonable formulation is suggested by the Turing machine.

Addition of two n-digit numbers can be performed on a Turing machine in a number of steps approximately proportional to n. More precisely, the number of steps divided by n approaches a finite limit as n approaches infinity. Let $M(n)$ denote the *minimal* number of steps it takes a Turing machine to multiply two n-digit numbers.

If one could prove that

$$\lim_{n \to \infty} \frac{M(n)}{n} = \infty$$

one could interpret this as meaning that multiplication is indeed more complex than addition.

Although this has not yet been proved, the closely related result that for every $\epsilon > 0$

$$\lim_{n \to \infty} \frac{M(n)}{n^{1 + \epsilon}} = 0$$

has been established. The interpretation of this is that after all multiplication is not much more complex than addition.

New problems call for new methods of attack, bring out new connections, and often reinforce old ones. In this way computing machines already have made significant contributions both to the *problematics* and to the *methodology* of mathematics. It is inconceivable that they will not continue to do so to an ever-increasing extent in spite of invective against them by some contemporary followers of Plato.

Outside of mathematics, electronic computers have proved to be indispensable tools.

It would take hundreds, perhaps thousands of pages, merely to list individual problems of mathematical physics whose solutions, effected on machines, have contributed to the examination of existing theories and to the suggestion of new properties of complicated physical systems. Especially in the new and rapidly developing areas of biology, computers are destined to play a very important role. Already, with their help, problems like that of the structure of certain organic molecules (*e.g.*, the location of atoms in the myoglobin molecule) have been solved. Mathematically, the problem involves unraveling the spatial configuration of an assembly of atoms from X-ray diffraction patterns presented by the whole molecule. Technically, the work involves, in part, inversion of Fourier transforms and manipulation of a great mass of statistical data. It is fair to say that a solution would have been impossible without modern computing tools.

In concluding this section, we should at least mention an important, though often overlooked, way in which the existence of computers tends to influence and polarize our thinking about a variety of problems.

Consider, for instance, the problem of machine translation from one language to another. Modern computers are entirely adequate to store vast dictionaries, and they permit retrieval at fantastic speed. But this is not enough without giving the machine at least modest instruction in the elements of grammar and syntax. This task of instructing an automaton calls for a sharply critical reexamination of the instructor's own linguistic knowledge. He cannot rely on a machine to have the complex, subtle psychological and cultural prerequisites that, perhaps unconsciously, he has come to expect from human beings. The problem he faces is profound; in essence, it is a question of how much of what we consider to be "intellect" can be traded for memory capacity and speed.

One may speculate that we are close to precise formulation of such problems and that computers will influence and perhaps even change our philosophy.

We have stressed repeatedly, though not always explicitly, that mathematics owes its unique position to its adherence to the axiomatic method.

As we have pointed out, this method consists in starting with a few statements (axioms) whose truth is taken for granted and then deriving other statements from them by the application of rules of logic *alone*.

The axioms are meant to describe simple properties of objects under consideration; one hopes that in these properties the essence of the objects will be captured completely. However, how does one know whether a system of axioms indeed has succeeded in capturing what was intended?

It is a little easier to decide when a system has failed in this task than when it has succeeded. It has failed if one can add to it a new and pertinent statement A or its negation $\sim A$ ("not A") and in each case obtain a system free from contradiction. In other words, a system of axioms is *not categorical* (*does not uniquely* characterize its objects) if there is a statement A that is *independent* of the axioms of the system (*i.e.,* if adding A or $\sim A$ to the system will not lead to a contradiction).

The question of the independence of axioms goes back to antiquity. The attempt to derive the axiom of parallels (Euclid's fifth postulate that through a point not on a line l there passes one and only one line parallel to l) from the remaining axioms was one of the strongest motives that drove the mathematics of the post-Euclid era. Failure to prove or disprove the fifth postulate despite enormous efforts in this direction by many generations of mathematicians no doubt was responsible for creating an aura of the absolute about Euclidean geometry. Kant, for example, considered Euclidean geometry as providing the only way in which people could deal with space in deductive terms, and he therefore sometimes is blamed for having delayed the discovery of non-Euclidean geometries.

The almost tragic character of the struggle against the fifth postulate is illustrated by the following passage from a letter the older Bolyai wrote to his son urging him to abandon his researches: "I have traveled past all reefs of this infernal Dead Sea and have always come back with broken mast and torn sail."

Even when Bolyai and Lobachevski independently had discovered the first non-Euclidean geometries, the logical status of the fifth postulate remained somewhat unclear.

Full clarification came a little later: first came the discovery of Beltrami that geometry on a certain surface (called a pseudosphere since it is a surface of *constant negative* curvature) provides a realization of the Bolyai-Lobachevski geometry; then Klein and Poincaré constructed extremely simple *plane* models in which the new geometry was valid.

All this led to the establishment of a general method for proofs of the independence of axioms. In brief, this method is as follows: To prove that an axiom A is independent of a system of axioms S one adjoins $\sim A$ (the negation of A) to S; then one tries to find objects that satisfy S and $\sim A$. If this can be done, the system consisting of S and $\sim A$ is clearly free from contradiction; thus A is independent of S, and A could be adjacent to S.

Usually in constructing objects that form a *model* for a set of axioms, one has to use another system in which freedom from contradiction is taken for granted. For example, plane analytic geometry provides a model for plane geometry by interpreting points as ordered pairs (x,y) of real numbers, straight lines as sets of pairs (x,y) satisfying a linear equation of the form $ax + by + c = 0$, etc. This interpretation of plane geometry depends on the system of real numbers. Certain logical difficulties notwithstanding, every mathematician proceeds on the assumption that there is no contradiction in the algebra of real numbers.

Here is how one proves the independence of the fifth postulate in plane geometry:

Consider (as in fig. 32) the upper half of the ordinary plane, excluding the x-axis.

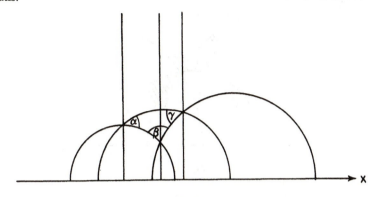

Fig. 32

The points of the new geometry simply will be points of this half-plane, but "straight lines" now will be interpreted as semicircles with centres on the x-axis (so that they cut the x-axis at right angles) or ordinary straight lines perpendicular to the x-axis. We define angles between our "straight lines" in the usual way (*i.e.*, as the *Euclidean* angles between tangents). To complete our description of this geometry we must define congruence, or equivalently, describe what are the "rigid motions."

The motions are one-to-one transformations of the half-plane into itself that transform "straight lines" (*i.e.*, semicircles with centres on the x-axis or ordinary straight lines perpendicular to the x-axis) into "straight lines" and that do not distort the angles (requirement of rigidity).

If one picks an origin on the x-axis and associates with each point P of the upper half-plane a complex number z

$$P \rightarrow z = x + iy \qquad y > 0$$

then it can be shown that the rigid motions are transformations $T(z)$ of the form

$$T(z) = \frac{\alpha z + \beta}{\gamma z + \delta}$$

where $\alpha, \beta, \gamma, \delta$ are *real* and $\alpha \delta - \beta \gamma > 0$.

The concept of betweenness is introduced in the obvious way, and one then can verify that *all* of Hilbert's axioms for plane Euclidean geometry (including the partly controversial axiom of continuity)[5] are satisfied *except* the axiom of parallels.

---

[5] As mentioned previously the axiom of continuity in effect establishes an order-preserving one-to-one correspondence between points of a straight line and real numbers. In the case of our concrete model such a correspondence can be easily and explicitly established.

As fig. 33 shows there is a whole *angle* (shaded in the figure) of lines through
P that are parallel to *l*.

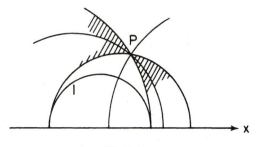

Fig. 33

That the result of centuries of striving can be summarized (and quite accurately) in so few words may come as a surprise. Bolyai and Lobachevski had
to work much harder because the concept of a model did not exist at the time.
They worked directly from axioms and used ordinary Euclidean drawings to
arrive at what must have appeared to them as highly bizarre conclusions; with
iron discipline they had to subjugate habit and intuition to the unbending de-
mands of logic.

Nowhere in mathematics are questions of completeness, categoricity, and independence of axioms more important than they are in set theory. Since set
theory is considered without dissension to be the basis of all of mathematics,
its axiomatic foundations are of paramount interest and importance.

Also in set theory we run into a proposition like the axiom of choice whose
consequences are so strange that some mathematicians prefer to avoid if not
to reject it outright.

Then there is Cantor's celebrated continuum hypothesis whose proof (or dis-
proof) was considered by Hilbert to be of such fundamental importance that
he placed it first on his celebrated list of unsolved problems.

Several years after publishing his paper on undecidable propositions, Gödel
showed that both the axiom of choice and the continuum hypothesis can be
assumed to be *true* in formal systems of set theory, like those of Fraenkel or Hilbert-Bernays. In other words he showed that they are either provable within
the system or else they are independent of the rest of the axioms and could
be added to the system if one wishes.

In 1960 came a definitive clarification of the status of the axiom of choice
and of the continuum hypothesis when Paul Cohen showed that in the usual
formal systems of set theory the axiom of choice and the continuum hypothesis,
without contradiction, might be assumed *not* to hold!

As mentioned above, the result of Godel was that neither the axiom of choice nor the continuum hypothesis are *demonstrably* false. This was done by constructing a *model* of set theory (satisfying, for example, the axioms of Fraenkel or of Hilbert-Bernays) such that either of these propositions or, for that matter, both of them are true within it. The existence of such a model does not mean that a proposition in question (*e.g.*, the continuum hypothesis) is provable from the axioms; just as constructing a model of a geometry that satisfies all the axioms of Euclid (including his fifth postulate) does not suffice to prove that postulate from the remaining axioms. What Paul Cohen did was to construct some *models* of set theory in which the axiom of choice does not hold and other models of set theory in which it does hold, but in which the continuum hypothesis does not. In fact, he exhibited a system that satisfies all the usual axioms of set theory, including the axiom of choice, but in which the continuum is of a "very high" power. Thus there *do* exist in the system sets of intermediate powers between that of $\aleph_0$ (*i.e.*, the power of the set of all integers), and that of $\mathfrak{c}$, the power of the continuum.

These results must be considered as definitely settling Cantor's problem, at least in the framework of present formulations of the theory of sets. In our view they present new challenges and open new perspectives in the foundations of mathematics. It should be borne in mind, however, that we repeatedly have used the phrase "formal axiomatic systems" for set theory. One must ask whether these systems are indeed sufficiently general to embody all our intuition; which in a sense is more fundamental than any codified set of expressions put on paper as supposedly reflecting this intuition. *In the present state* of mathematical thinking this question must be considered as belonging really to metamathematics. This, however, is of great philosophical importance; as we have seen on previous occasions, throughout the history of our science, ideas that originally were metamathematical eventually became parts of mathematics itself. Metamathematics is of considerable potential mathematical interest.

It may be that, in the future, no finite system of axioms will be considered definite or ultimate; but, almost in analogy with the world of living entities, new axioms slowly will be added. This will be through a consensus among mathematicians that will arise (as it were, genetically) from previous axioms and from the experiences that the consequences of the axioms will bring about. It also could be argued that no formal system considered up to now adequately embodies the image of the infinite that is unconsciously held by mathematicians; one might even venture a conjecture that no such *formal* system is possible.

In discussing themes and trends of mathematics we have stressed questions of foundations, especially those of set theory; but we do not want to leave the impression that the main body of mathematics was directly affected by the profound clarifications that have resulted from the work of Gödel or Paul Cohen.

In fact, most of mathematics was not affected at all. On the other hand, the clarification of the foundations of geometry resulting from the discovery of non-Euclidean geometries has had a profound effect on the main body of mathematics and on physics and astronomy as well.

Perhaps this is so because the logical and set-theoretic foundations of mathematics are too all-embracing and of too general a character to play a vital role in specific problems of everyday mathematics. Perhaps it is because they deal more with the process of deduction than with its fruits.

Be this as it may, work in the foundations of mathematics as a *whole* has yielded a negative result, for it has underlined the limitations of the axiomatic method. In set theory, it actually produced serious doubts as to whether there are formal systems capable of describing what working mathematicians feel sets ought to be.

Work on the axiomatic foundations of geometry, on the other hand, had a different and a much more constructive effect. It turned out that much of our intuition about space can be deductively codified. Also the codification contained in itself seeds of new worlds like those of Bolyai and Lobachevski, worlds that at first went beyond physical intuition only to merge with it later in the context of relativity theory. Last but not least, this work inspired important developments in differential geometry, a central field of mathematics that is still vigorously studied. One is almost tempted to conclude that in some undefined way there is a deep distinction between the problem of axiomatizing a branch of mathematics that originated in *external* stimuli and the problem of axiomatizing *internal* processes of thinking.

Chapter 3   Relations to Other Disciplines

THE RELATIONS BETWEEN MATHEMATICS and the empirical disciplines have puzzled philosophers and historians of science for centuries.

There is little doubt that the "external world" has been the source of many mathematical concepts and theories. But, once conceived, these concepts and theories evolved quite independently of their origins. More often than not, they soared to heights of abstraction that left behind their concrete (and perhaps even humble) backgrounds. In this evolutionary process, new concepts and theories were generated through highly introspective activities that, in turn, frequently had miraculous and decisive influence on scientific developments outside of mathematics proper.

As an example take geometry.

Originating in geodesy and astronomy, its first great period of growth culminated in Euclid's *Elements,* which for centuries served as an unsurpassed model of logical perfection and purity.

In the self-imposed isolation from the external world that first inspired its creation, geometry continued to grow by feeding on its own problems. Of these, the problem of the fifth postulate (discussed in Chapter 2) was as elusive as it was fascinating.

It was a purely logical problem concerning, as we have seen, the question of whether the axiom (postulate) can be derived from the other axioms.

That the answer was in the negative was first established by Bolyai and Lobachevski by exhibiting a system of geometrical propositions (including a negation of the fifth postulate) that were in such a one-to-one correspondence with their Euclidean counterparts that a contradiction in one system would immediately imply a contradiction in the other.

It is interesting to note that neither Bolyai nor Lobachevski had a clear realization of the "reality" of their geometry. Lobachevski called his geometry "imaginary" and Bolyai, in a moving letter to his father, wrote that "out of nothing I have created a new and wonderful world."

It was many years later that Bolyai-Lobachevski geometry helped Riemann to propose a deep and far-reaching approach to non-Euclidean geometries. The resulting mathematical apparatus became the foundation of Einstein's general theory of relativity.

The example of geometry is perhaps the most dramatic, but it is far from unique in illustrating the transmutations of mathematical concepts and ideas.

While considering how the earth cooled, Fourier was led to the problem of representing periodic functions as series of sines and cosines of the form

$$\tfrac{1}{2}a_0 + \sum_{n=1}^{\infty}(a_n \cos 2\pi nx + b_n \sin 2\pi nx)$$

One is led to the same problem in attempting to resolve a periodic disturbance (*e.g.*, a sound wave produced by a musical instrument) into "pure tones" (sinusoidal disturbances).

These physical problems provided a strong impetus for the study of series of sines and cosines like the one above, and it led to a purely mathematical theory of trigonometric series.

As the theory grew it became apparent that parts of it are quite independent of the special nature of sinusoidal "pure tones." In fact, a large body of theory still could be retained when the very special, though physically appropriate, sines and cosines were replaced by functions $\phi_n(x)$, subject only to the condition that

$$\int_0^1 \phi_n(x)\phi_m(x)dx = \begin{cases} 0, & m \neq n \\ 1, & m = n \end{cases}$$

This condition is analogous to the condition of perpendicularity of vectors in Euclidean space (see Section 13 of Chapter 1) combined with the condition that the vectors be of unit length. In this way the problem of representing a function as a series

$$\sum_{n=1}^{\infty} c_n\phi_n(x)$$

became analogous to resolving vectors into mutually perpendicular components.

This and other closely related analogies readily led to the introduction of the simplest infinite-dimensional space, the so-called Hilbert space. Then, again miraculously, Hilbert space provided the proper mathematical framework for quantum mechanics.

It is well known that the development of mathematics was strongly, and, at times, decisively, influenced by problems of physics and astronomy.

Infinitesimal calculus, perhaps still the greatest single step in the evolution of mathematical concepts and methods, was developed by Newton to deal with problems of dynamics, especially those posed by the motion of planets. The crowning achievement of Newton's work was the derivation of Kepler's laws [1] from the law of gravity.

---

[1] These laws are: (i) Planets move in confocal ellipses around the sun which is situated at one of the foci. (ii) The line (radius vector) connecting the sun and a planet sweeps out equal areas in equal

The law of gravity states that two bodies are attracted with a force proportional to their masses and inversely proportional to the square of the distance between them. Once postulated, Newton's second law of motion (*i.e.*, that force is equal to mass times acceleration) can be used to conclude that the acceleration of a planet is inversely proportional to the square of its distance from the sun and is directed toward the sun along the line connecting the sun with the planet.

Since the acceleration is the second derivative of the radius vector of the planet, one thus obtains an equation that connects the second derivative of a vector with the vector itself. It is called a differential equation since it involves a derivative of the radius vector whose dependence on time is what one seeks. It is this equation that Newton derived and solved, obtaining as a consequence all three of Kepler's laws.

It is difficult to convey the impact of this great feat on the course of science. It certainly was the beginning of theoretical physics as we now know it, and it established a pattern for using mathematical concepts in the description of physical events.

The basic operations of calculus, differentiation and integration, proved entirely sufficient to formulate all the physical laws discovered during the 18th and 19th centuries.

Theories of elasticity and fluid flow, thermodynamics, and Maxwell's great theory of electromagnetic phenomena all were tributes to the almost miraculous versatility of the infinitesimal calculus.

No wonder that analysis, the part of mathematics that grew out of the ideas of calculus, became *the* language of the exact sciences and that mathematicians could proudly participate in the assault on the mysteries of nature.

During the past two centuries, physics has been very mathematical and mathematics has been involved closely with and influenced heavily by physics. Many great figures in mathematics of that period also were leading physicists. The tradition of close cooperation between the two disciplines continues to this day, though on a greatly reduced scale.

How fruitful and far-reaching the consequences of such cooperation can be is illustrated by the prediction of electromagnetic waves and by the electromagnetic theory of light. By the middle of the 19th century there was a large body of experimental material concerning electromagnetic phenomena. On the basis of this material, Maxwell, by a combination of deduction and daring, proposed a set of differential equations that implied and codified what was known about electricity and magnetism at that time.

---

intervals of time (this has the effect that, *e.g.*, the earth moves faster in the winter when it is closer to the sun). (iii) The *squares* of planetary years (*i.e.*, the time it takes any planet to go around the sun once) are in the same ratio as the *cubes* of their distances from the sun.

In a vacuum, electromagnetic phenomena are such that Maxwell's equations become particularly simple and contain only two vectorial quantities: the electric field $\vec{E}$ and the magnetic field $\vec{B}$. The equations are:

$$\nabla \cdot \vec{E} = 0, \ \nabla \cdot \vec{B} = 0$$

$$\alpha \frac{\partial \vec{E}}{\partial t} = \nabla \times \vec{B}, \ \beta \frac{\partial \vec{B}}{\partial t} = - \nabla \times \vec{E}$$

The scalar quantities α and β depend on the choice of units. From these equations one can derive by straightforward *mathematical* manipulations that the electric field $\vec{E}$ satisfies the equation

$$\frac{\partial^2 \vec{E}}{\partial t^2} = \frac{1}{\alpha\beta} \nabla \cdot \nabla \vec{E}$$

$$= \frac{1}{\alpha\beta}\left(\frac{\partial^2 \vec{E}}{\partial x^2} + \frac{\partial^2 \vec{E}}{\partial y^2} + \frac{\partial^2 \vec{E}}{\partial z^2}\right)$$

and that the magnetic field $\vec{B}$ satisfies the same equation; *i.e.*,

$$\frac{\partial^2 \vec{B}}{\partial t^2} = \frac{1}{\alpha\beta} \nabla \cdot \nabla \vec{B}$$

$$= \frac{1}{\alpha\beta}\left(\frac{\partial^2 \vec{B}}{\partial x^2} + \frac{\partial^2 \vec{B}}{\partial y^2} + \frac{\partial^2 \vec{B}}{\partial z^2}\right)$$

The quantity $1/\alpha\beta$ has the dimension of the *square of a velocity* and can be determined experimentally. The remarkable result is that

$$\frac{1}{\alpha\beta} = c^2$$

where c is the velocity of light!

Now, it had been known for quite some time before Maxwell that a local disturbance in an isotropic elastic medium (initially at rest) propagates as waves governed by the *wave equation*

$$\frac{\partial^2 U}{\partial t^2} = c^2\left(\frac{\partial^2 U}{\partial x^2} + \frac{\partial^2 U}{\partial y^2} + \frac{\partial^2 U}{\partial z^2}\right)$$

where $U(x,y,z,t)$ is the displacement from the initial position of rest at the point (x,y,z) at time t. The constant c is the velocity at which the waves propagate through the medium.

Maxwell was struck by the fact that the electric and magnetic vectors obey the wave equation, and he concluded that electromagnetic disturbances propagate as waves. This brilliant theoretical prediction was dramatically confirmed when in 1886 Heinrich Hertz experimentally produced electromagnetic waves. Since electromagnetic waves propagate with the velocity of light, Maxwell also proposed the theory that light is a form of electromagnetic radiation. This, too, was fully confirmed by numerous experiments and further theoretical considerations, and a profound insight into the nature of light was thus gained.

Maxwell's theory of electromagnetic phenomena also can be used to illustrate another way (in a sense more subtle) in which mathematical and physical ideas are brought together. It concerns the fact that Maxwell's equations are *not* invariant under the Galilean transformations (see Section 14 of Chapter 1) while the laws of Newtonian dynamics are.

On the other hand, Maxwell's equations are unchanged by the Lorentz transformations (see again Section 14, Chapter 1). This is a purely mathematical fact implied by the *form* of Maxwell equations that in principle could be discovered without knowledge or understanding of the physical content of the equations. The bold leap of requiring that equations of dynamics then be modified suitably to make them also invariant under the Lorentz transformations is no longer mathematical or even deductive. It is *an* answer to the dilemma posed by the negative outcome of the Michelson-Morley experiment (Section 14, Chapter 1); it implies that all laws of physics should be invariant under the transformations of the Lorentz group.

When Einstein formulated this far-reaching view of physics in 1905, the ideas of Felix Klein on the nature of geometry were widely known and fully appreciated by the mathematicians of the day. Klein summarized these ideas in his inaugural address as professor of mathematics at Erlangen. The address became known as the Erlangen Program, and considered geometries as studies of *invariants* of appropriate groups of transformations. The great mathematician Hermann Minkowski was impressed by the *conceptual* similarity between the ideas of Einstein in physics and those of Felix Klein in geometry. Minkowski went on to produce a beautiful blend of the two lines of thinking by the introduction of space-time endowed with a geometry based on the Lorentz transformations.

In discussing the role mathematics plays in the formulation of physical laws and in drawing conclusions from these laws, mention should be made of the frequent discrepancy between the depth of physical insight and the complexity of the corresponding mathematical description.

The *technical* mathematical apparatus of the special theory of relativity is elementary in the extreme; the underlying physical concepts and ideas are subtle and deep. By way of contrast, many problems posed by technology contribute little to our understanding of the physical world, although they call for the use of extremely complex mathematical techniques. Also, while it is remarkable that mathematics conceived and nurtured internally quite frequently finds unexpected uses in symbolic description of external phenomena (complex numbers and matrices are good examples), neither elegance nor intricacy is in itself a guarantee that a mathematical concept, construct, or method will prove empirically relevant or useful.

E. P. Wigner summed this up in writing on "The Unreasonable Effectiveness of Mathematics in the Natural Sciences":

> The miracle of the appropriateness of the language of mathematics for the formulation of the laws of physics is a wonderful gift which we neither understand nor deserve. We should be grateful for it and hope that it will remain valid in future research and that it will extend, for better or for worse, to our pleasure even though perhaps also to our bafflement, to wide branches of learning.[2]

It would be futile to attempt anything approaching a complete coverage of interaction between mathematics and the physical sciences. However, one kind of interaction that is likely to be overlooked is of considerable interest and importance; it is the following:

The external world is complex, and the natural scientist is gratified merely to perceive and understand some of its simplest properties. To do so he sets up simplified and idealized *models* that have, he hopes, the essential properties of physical objects stripped of irksome and less-important complications.

Thus Newton explained Kepler's laws of planetary motion by treating each planet as subject only to the gravitational force of the sun. He neglected the pull of all other heavenly bodies even though, strictly speaking, it was wrong to do so. Later, more realistic models were introduced. In fact, one of the great feats of 19th-century astronomy was the prediction by Adams and Leverrier of the existence of the planet Neptune through attempts to account for the relatively sizable deviations of the motion of Uranus from its Keplerian orbit.

Roughly speaking, it is the natural scientist who decides on a model; mathematics then comes in by drawing conclusions (deductively) from the model. This pattern is so well known that it hardly requires elaboration.

There are other kinds of models suggested by logical and conceptual difficulties in dealing with seemingly sound models suggested by external phenomena. As an example consider two bodies A and B brought in thermal contact and isolated from all other bodies. Thermodynamics then predicts a *unidirectional* flow of heat from the warmer body (say A) to the cooler one (B), the temperature difference tending exponentially toward equalization (Newton's Law of Cooling). This is a consequence of the famed Second Law of Thermodynamics which, in its most pessimistic form, predicts eventual total equalization, what Clausius called *Wärmetod* ("heat death").

The mechanistic (kinetic) view, picturing matter as composed of particles, namely atoms or molecules, obeying the usual laws of dynamics, leads to an entirely different picture. The particles, bouncing against each other and moving in what appears to be a "random" fashion, surely will not produce an *ab-*

[2] *Communication in Pure and Applied Mathematics*, vol. 13, Courant Institute of New York University, New York, 1960, pp. 1–14.

solutely unidirectional flow from A to B. As a matter of fact, by a theorem of Poincaré, such a dynamic system eventually will return to a state arbitrarily near its initial condition unless it starts from such an exceptional state that the possibility can be safely neglected. This "quasiperiodic" behaviour of dynamic systems contrasts sharply with the monotonic trend toward equalization implied by the Second Law.

To clarify the issues involved, Paul and Tatiana Ehrenfest proposed in 1907 a simple and beautiful model (briefly described in Section 16 of Chapter 1).

Consider two boxes A and B, one of which (say A) contains initially a large number N of numbered balls (in Section 16, Chapter 1, N was taken to be $2R$). We now play the following game: We choose "at random" a number between 1 and N and move the ball of that number to the other box; the first move, of course, is from A to B. The process is then repeated many times (often returning balls to A), consecutive drawings being independent and all numbers from 1 to N being equally likely.

Intuitively, as long as there are many more balls in A than in B, the probability of moving from A to B will be correspondingly greater. We thus can expect a sort of unidirectional flow from A to B.

Although the drawing of numbers is independent, the quantities of balls in A in consecutive instances are not. They exhibit a kind of dependence called a *Markov chain* (see Section 16, Chapter 1). One finds that the *average number* of balls in A decreases exponentially to $N/2$, a result in complete agreement with the thermodynamic prediction. One also finds that *with probability equal to unity the model eventually will return to its initial state* (*i.e.*, all balls in A). This is the counterpart of Poincaré's theorem about dynamic systems.

Evidently there is no real contradiction between the Second Law and the inherently quasiperiodic behaviour of dynamic systems, once we give up the absolute dogmatism of the Second Law and allow a more flexible interpretation based on probability theory.

All this would be reinforced if one were to calculate how long on the average one would have to wait for the return of the initial state in the Ehrenfest model. The answer is 2^N steps, which is staggeringly large even for moderate N, say, about 100.

If we seem to observe all around us irreversible (unidirectional) phenomena, it is simply that our life span is so pitifully short compared with those enormous times of return!

With modern computers it is easy to play the Ehrenfest "game." Experiments have been performed with $N = 2^{14} = 16,384$ "balls," each run consisting of 200,000 drawings. (It takes less than two minutes.) The number of balls in A was recorded after every 1,000 drawings and one of the resulting graphs is shown as fig. 34.

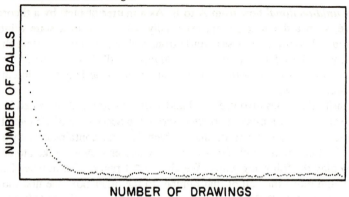

Fig. 34. Played on a computer, an Ehrenfest game with 16,384 hypothetical balls and 200,000 drawings took just two minutes. Starting with all the balls in container *A*, the number of balls in *A* was recorded with a dot every 1,000 drawings. It declined exponentially until equilibrium was reached with 8,192 balls (half of them) in each container. After that fluctuations though small are quite nearly visible.

As one can see, at first the number of balls in A falls off along a nearly perfect exponential. But near equilibrium the graph gets a bit "wiggly," indicating the presence of fluctuations.

As a model of heat equalization, the Ehrenfest model is far from the reality of the phenomenon itself. It nevertheless captures the essence of the reconciliation between the kinetic view of matter and traditional thermodynamics.

During the 20th century, mathematical concepts, methods, and techniques were beginning to permeate more and more areas of learning and application. One might even venture the statement that we are witnessing a trend toward the "mathematization" of the bulk of intellectual activity. Not in all respects is this trend justified. One can point to numerous instances in which "mathematization" is trivial or pretentious, and even to some where it is both.

Taste or judgment notwithstanding, there is no denying that an ever-increasing number and variety of problems have become amenable to a formulation and treatment that is *par excellence* mathematical. Of these, we single out for a brief discussion just three: the theory of queues, the theory of games, and information theory.

The theory of queues originated in attempts to design a central telephone exchange that in some way would minimize the inconvenience of waiting for a connection. The simplest type of problem in this regard is the following:

Suppose that "customers" (these may be telephone calls) arrive to be served (or processed) at a station with one server and that they are served in series, one after another. Also suppose that time can be subdivided into elementary intervals of duration τ. It is not necessary to "quantize" time, but it is easier to state the problem if one does. Ultimately one can let τ approach zero in an appropriate way and obtain a theory in which arrivals are continuous in time. Then suppose that the probability that k (where $k = 0,1,2\ldots$) customers will show up during a given interval is p_k.

This means that

$$p_0 + p_1 + p_2 + \cdots = 1$$

A far-reaching simplification is then introduced by assuming that arrivals in distinct time intervals are *independent events* so that the probability that k_1 customers arrive during the first time interval, k_2 during the second, k_3 during the third, etc., is the product

$$p_{k_1} p_{k_2} p_{k_3} \cdots$$

Finally, it is assumed that the service time is governed by chance, and the probability that it takes time $\lambda\tau$ (*i.e.*, λ elementary time intervals where $\lambda = 1,2,3,\ldots$) to complete the serving process is assumed to be ρ_λ. This means that

$$\rho_1 + \rho_2 + \rho_3 + \cdots = 1$$

One now can ask a number of pertinent questions: What is the average number of customers waiting to be served after a prescribed time has elapsed? What is the average time that a customer has to wait before he is served? These questions can be fully answered, but the answers are by no means simple. The road leading to them passes through unexpected mathematical areas.

For example, one is led to consider power series

and
$$p_0 + p_1 z + p_2 z^2 + \cdots$$
$$\rho_1 w + \rho_2 w^2 + \rho_3 w^3 + \cdots$$

for *complex* values of z and w; as a result the theory of functions of a *complex variable* plays a decisive role.

As soon as the model is made more realistic by allowing more than one server, the mathematical difficulties become almost insurmountable, and even the simplest questions cannot be answered with satisfactory completeness. Fortunately, high-speed computers come to the rescue of the designer of a complex multiserver system like a telephone exchange. By adroit use of the Monte Carlo method (described in Chapter 2) one can simulate a projected system and empirically investigate various aspects of its operation.

Strictly speaking, such an "experimental" approach is not a part of mathematics. But the question is not unlike that posed centuries ago by the attempts of Eudoxus and Archytas noted earlier to inject mechanics into the mainstream of geometry (p. 133).

It is quite likely that empirical studies of queues in complex systems will suggest an analytic attack that will call for new concepts and techniques. These in turn may well influence and enrich distant and unrelated parts of mathematics.

The theory of games, which was invented almost single-handedly by John von Neumann, provides a striking example of how one can "mathematize" a body of problems that seemingly are outside the reach of any kind of rational approach. We shall explain what the theory is about in an example [3] of a highly simplified game of poker.

Suppose we have a deck of $2n$ (n very large) cards, half of which are marked H (high) and the other half marked L (low). Two players A and B "ante" an amount a, and are dealt one card each. Now, A can either "see" (*i.e.,* demand that B expose his card), or "raise" by an amount b (*i.e.,* put an additional b dollars into the "pot"). If A chooses "see," B has no choice but to expose his card, and he wins a (A's ante) if his card is H while A's is L, and loses a (his own ante) in the opposite case. The pot is split if the cards are both H or both L.

However, if A "raises," B has a choice either to "fold" (*i.e.,* withdraw and let A win a), or to "call" (*i.e.,* put in the amount b and thereby force A to expose his card). Again winning, losing, or splitting the pot depends in the obvious way on the values of the cards held.

The problem now is how should A and B play to their respective advantages? A "pure strategy" is a rule that tells the player what to do in each situation he may encounter. Player A has therefore four pure strategies available to him:

(1) (S,S) a "see-see" strategy; *i.e.,* "see" regardless of whether he has been dealt H or L.

(2) (S,R) or "see-raise" strategy; *i.e.,* "see" if he has been dealt H and "raise" if he has been dealt L.

(3) (R,S) or "raise-see" strategy; *i.e.,* "raise" on H and "see" on L.

(4) (R,R) a "raise-raise" strategy; *i.e.,* "raise" regardless of the card dealt.

One should note that (S,S) and especially (S,R) are not "good" strategies because neither takes advantage of the good fortune of having been dealt a high card.

Similarly player B also can choose from among the four "call" (C) or "fold" (F) strategies: (F,F), (F,C), (C,F), and (C,C). It should be recalled that when A elects to "see," B has no choice. Of these the first and second strategies require B to "fold" with a high card and therefore they are not "good" for B.

If we assume that A and B play the game to enrich themselves, rather than to practise disguised charity, we can disregard outright (S,S) and (S,R) for A and (F,F) and (F,C) for B. It is now easy to calculate how much A stands to gain for each combination of his own and B's strategies.

---

[3] This example is due to A. W. Tucker and is a simplified variant of one by von Neumann and Morgenstern.

Suppose, for example, that A chooses (R,S) ("raise" on H, "see" on L) and B chooses (C,C) ("call" on both H and L). The following table is nearly self-explanatory.

A is dealt	B is dealt	A wins
H	H	0
H	L	$a + b$
L	H	$- a$
L	L	0

If n is very large, the combinations (H,H), (H,L), (L,H), and (L,L) each will appear with the same approximate frequency; namely ¼. On the "average," then, (R,S) versus (C,C) strategies will net A a gain of $b/4$ dollars per game.

Similarly, one determines the average net gains per game for A in the remaining three strategy confrontations, and the results are displayed in the form of a so-called payoff matrix.

A \ B	(C,F)	(C,C)
(R,S)	0	$\dfrac{b}{4}$
(R,R)	$\dfrac{a - b}{4}$	0

Suppose that $a < b$ so that the left lower corner entry in the payoff matrix is negative. Then surely (R,R) strategy is disadvantageous to A, and he will choose the pure strategy (R,S). Similarly the (C,C) strategy is disadvantageous to B, and he will choose the pure strategy (C,F). Thus, by playing conservatively ("raising" with H, "seeing" with L; "calling" a "raise" with H, "folding" with L) both players can be *assured* of staying even with the board on the average. The optimal strategies are pure, and the game is fair.

If $a > b$, the game is biased in favour of A because only A has the privilege of "raising"; in actual practice this privilege is rotated among the players. However, A will lose the advantage of the bias unless he decides on a *mixed strategy* of choosing (R,S) with probability p_1 and (R,R) with probability p_2, where $p_1 + p_2 = 1$.

For example, if $a = 8$ and $b = 4$, the payoff matrix is

$$\begin{pmatrix} 0 & 1 \\ 1 & 0 \end{pmatrix}$$

and A can *assure* himself of an average gain of ½ per game by choosing (R,S) and (R,R), each with probability of 50%. By the same token B, by choosing his strategy on a fifty-fifty basis, can *prevent* A from averaging more than ½ per game.

We thus see that in some situations A must *"bluff"* (*i.e.*, "raise" with a low card) part of the time to achieve an optimal result; precisely what part is determined by the payoff matrix.

Von Neumann has shown that a large class of competitive encounters, like those arising in economics, can be formulated in terms of *matrix games; i.e.*, games based on an n by m payoff matrix

$$\begin{pmatrix} a_{11} & a_{12} & \ldots & a_{1n} \\ a_{21} & a_{22} & \ldots & a_{2n} \\ \ldots\ldots\ldots\ldots\ldots\ldots\ldots \\ a_{m1} & a_{m2} & \cdots & a_{mn} \end{pmatrix}$$

in which a player A chooses a row i and "wins" a_{ij} when his competition B chooses (unknown to A!) the column j.

The fundamental theorem of game theory asserts that there is a number v, called the value of the game, such that A can *assure* himself of winning on the average at least v per game, while B can *prevent* him from gaining more than that. Moreover, A has an optimal strategy (in general, mixed) that will guarantee him at least v per game, and B has an optimal strategy (also in general, mixed) that will guarantee him that he will not lose more than v.

It is perhaps too early to judge the results of applying game theory, especially in economics where it has found some of its best-known (and best-advertised) uses. For one thing, because of the large size of payoff matrices in realistic situations, a complete numerical analysis is still almost out of reach even with high-speed computers.

Aside from its specific applicability or usefulness in this or that area of knowledge, game theory has played an important part in bringing mathematical thinking to bear on a set of questions and problems relating to what might be termed rational behaviour in competitive situations. Even if its models have been so oversimplified as to be wholly unrealistic, game theory deserves great credit for offering hope for a disciplined approach to enormously complex problems involving social behaviour.

One cannot leave the subject without at least brief mention of Abraham Wald's statistical decision theory, which was inspired by game theory. Wald envisaged the process of making a decision in the presence of chance as a game between the statistician and nature. Nature's strategy is not known, of course, but the statistician makes his decisions according to an optimal strategy dictated by a payoff matrix. The payoff matrix is set up according to an evaluation by the statistician of the relative cost to himself of different decisions. Formally, the theory is analogous to game theory, but since payoff matrices are now in most cases infinite, it is technically much more difficult and sophisticated. The impact of decision theory on statistics has been mainly conceptual. It has clarified and brought into sharp focus many of the basic questions of statistical inference, especially those related to the nature of statistical tests.

Information theory deals with problems connected with the efficient transmission of messages.

In a typical situation we have an information source that selects *one message* to be sent from a set of messages, a transmitter that changes the message into a *signal,* an appropriate channel through which the signal is sent, and a receiver that changes the signal back into a message. For example, in telegraphy, written words are coded into sequences of interrupted currents of varying length (dashes, dots, spaces) and sent over a wire to be reconverted into written words.

In information theory one is not concerned with the semantic problem of how well the transmitted symbols convey the desired meaning but only with the technical problem of accurate and economic transmittal.

To explain and illustrate the kind of questions that arise in information theory, suppose that the messages are simply strings of N (N large) letters of the Roman alphabet, each letter occurring with the same frequency as it does in English.

We may as well think of a somewhat more general situation in which we have an alphabet of k letters $S_1, S_2, \ldots, S_k$ where S_1 is to be chosen with probability p_1, S_2 with probability p_2, etc. The successive letters are chosen independently. If we now want to transmit such messages we could do it straightforwardly, letter by letter. Assuming that it takes some unit of time (*e.g.,* one microsecond) to transmit a letter, we see that the rate of transmission is one symbol per time unit. Can one do better? In fact, what is the minimal rate at which a transmission can be accomplished?

Transmission letter by letter is inefficient, for we do not take advantage of the fact that some messages are much less likely to be selected by the source than others. We therefore might be able to increase the rate of transmission by assigning short code names to frequent messages and reserving long code names to the infrequent ones.

Shannon has shown how to define a quantity H, called the entropy of the source, and another quantity C, called the capacity of the channel, and has then proved that the optimal transmission rate is C/H, which is greater than or equal to 1. This means that one can invent codes that will allow transmission at any average rate that is less than C/H, but there is no possible code that will allow the average rate of transmission to exceed C/H.

The entropy H is defined (roughly speaking) as $-\dfrac{1}{N} \times$ logarithm of the probability of a "typical" message. The capacity C is the maximum value of H over all possible assignments of probabilities that are consistent with constraints placed upon the messages. In the simple case under consideration, strings of N letters chosen independently, there are no constraints on the source. We shall come back to the matter of constraints a little later on.

For theoretical purposes it is sufficient to consider only *binary codes* (*i.e.*, codes which are sequences of 0's and 1's); so it is convenient to take 2 as the base for all logarithms in information theory. This merely amounts to a choice of units and is a matter of convention rather than of necessity.

To get a feeling for what is meant by a "typical" message, we go back to our simple example.

If N is large, most messages will contain *approximately* p_1N S_1's, p_2N S_2's, etc. This is a rough statement of a law of large numbers discussed in Chapter 1. A typical message *actually* contains p_1N S_1's, p_2N S_2's, etc. Of course, p_1N, p_2N, etc., need not be integers and should be replaced by integers nearest to them; but this makes no difference at all in the limit of large N (*i.e.*, $N \to \infty$).

The probability that a message of N letters will contain p_1N S_1's, p_2N S_2's, etc., is

$$p_1^{p_1N} p_2^{p_2N} \cdots p_k^{p_kN}$$

and we see therefore that in our primitive example the entropy is given by the formula

$$H = - (p_1\log p_1 + p_2\log p_2 + \cdots + p_k\log p_k)$$

The only restriction is that

$$p_1 + p_2 + \cdots + p_k = 1$$

It is seen that the maximum possible H when the p's are so restricted is

$$H_{\max} = C = - \log k$$

which occurs when all the p's have the same value. Thus one can encode to achieve any transmission rate less than

$$\frac{\log k}{p_1\log p_1 + \cdots + p_k\log p_k}$$

Up to now, except for prescribing frequencies of individual symbols, we have placed no constraints on the messages. On the other hand, a language like

English or French places severe constraints on sequences of letters that constitute allowable messages; not all of these constraints are known explicitly. One can approximate a real language by imposing more and more constraints of a statistical nature on the process of generating messages. For example, rather than choosing the letters independently we can insist that each digram (*i.e.,* each *pair* of successive letters) appear with the same frequency as it appears in the language, thus coming closer to the actual linguistic structure. One can go on in this way by adjusting frequencies of triples, quadruples, etc., of successive letters. If one is content with digrams, one achieves a statistical description of the source in terms of a simple Markov chain.

On our own artificial example based on the alphabet $S_1, S_2, \ldots, S_k$, it means that we are given probabilities p_{ij} that S_i will be followed by S_j and that the probability of the message

$$S_{i_1} S_{i_2} \ldots S_{i_N}$$

is

$$p_{i_1 i_2} p_{i_2 i_3} \cdots p_{i_{N-1} i_N}$$

The probabilities $p_1, p_2, \ldots, p_k$ with which individual symbols appear in long messages can be found by solving the linear equation

$$\sum_{i-1}^{k} p_i p_{ij} = p_j \qquad (j = 1, 2, \ldots, k)$$

One also can determine the entropy of this source, the result being

$$H = -\sum_i p_i \sum_j p_{ij} \log p_{ij}$$

One can show that this is *less* than or equal to $-\sum p_i \log p_i$, which would have been the entropy of the source if the symbols were generated independently with probabilities p_i. This is a special case of the general principle that more structure implies less entropy.

If some p_{ij} is either 0 or 1 (*e.g.,* in English the letter z is *never* followed by x so that $p_{zx} = 0$), we have an *absolute* constraint. In maximizing the entropy we can vary all p_{ij}'s except those that are either 0 or 1.

So far we have assumed that the channel is noiseless; *i.e.,* that every symbol is transmitted with absolute accuracy. The most interesting and mathematically most challenging problems arise when the channel is "noisy." The simplest model of such a noisy channel is a binary channel without memory. This means that in transmitting binary codes there is a constant probability p that the symbol 0 or 1 will be transmitted correctly and a constant probability $q = 1 - p$ that it will be "garbled" in transmission (0 changed into 1 or 1 into 0); furthermore, individual symbols are transmitted independently.

Shannon and others have shown how and under what circumstances one can construct codes that can be deciphered with arbitrarily high probability. Optimal rates of transmission also have been determined.

These more advanced developments are quite intricate; but even our brief and incomplete sketch of the more elementary parts of the theory should show how incisively and fruitfully mathematics is used today for problems that only a short time ago were considered beyond quantitative and deductive discussion.

No discussion of the relations of mathematics to nonmathematical disciplines can ignore statistics. Statistics is *not* a branch of mathematics since it is concerned with processing data and with making decisions based on the results of such processing. So used, it is not even a properly circumscribed discipline, but rather a general instrument of scientific methodology. However, mathematics has played and continues to play an important role in the growth and the development of statistics. In fact, a significant part of statistics has become so permeated with mathematical ideas and techniques that it is called mathematical statistics. In return, the statistical point of view has influenced many branches of pure mathematics by extending their problematics and by suggesting novel ways of approach.

It should perhaps be stressed again that the boundaries between mathematics and the many disciplines to which it is applied are seldom sharply drawn. Nothing but impoverishment can be expected from unfortunately rather frequent current efforts to isolate a body of "pure" mathematics from the rest of scientific endeavour and to let it feed only on itself.

Chapter 4 Summary and Outlook

OUR PEREGRINATIONS through mathematics have been guided by the historical development, inner connections, and growth of synthetic patterns of thinking in our science. We started with problems about integers in which ideas of infinity appeared, and proceeded to examples from geometry through the evolution of more abstract ideas about numbers and geometrical objects. We have attempted to show in an elementary way how mathematicians came to consider groups of general transformations and then, looking upon the sets of such objects as spaces, how they attempted to build theories of structures in general. Although the mathematical objects are today enormously varied, the mathematical method remains the same. A small number of axioms is postulated or implied. Then through repeated applications of a well-defined set of rules (mathematical logic) one builds theories; *i.e.,* collections of theorems describing the properties of and relations among objects that satisfy the axioms. Thus, to mathematicians like Archimedes, Euclid, or Newton, could they come alive now, the variety of notions that interest mathematicians might seem bewildering; but the methods would appear entirely understandable and familiar.

In our short account we have had to restrict ourselves to a selection of topics and to a selection of methods presently employed in mathematical research and in applications of mathematics. Some very recent modern techniques could not even be mentioned or hinted at. We now wish to identify some areas of particularly active mathematical research and to indicate how the sciences and technology are increasingly being mathematized.

The mathematical method has had its great triumphs in abstracting certain properties of observed facts and observed relations between them and obtaining through purely logical processes new relations which could then be verified by observation and by experimentation. Thus, Newton's formulations of the laws of dynamics made it possible to erect the edifice of classical mechanics purely through mathematics. The motions of celestial bodies were explained on this basis. In this spirit mathematical physics has had its continued successes, making possible the development of other sciences and of new technologies. In mathematics itself new theories are built by assuming a few mathematical properties. The theory of probability as discussed earlier is an example, as are theories

161

of geometries, analytic functions, and spaces whose elements or "points" are themselves functions. This process of axiomatic buildup of new domains continues. From observed classes of phenomena we abstract simpler or more "fundamental" classes, postulate a few of their properties, and then draw mathematical conclusions from models thus generated. In the same way, mathematical classes are studied through introspection, the effort being to unify these under the mantle of new "super theories." In other words, new mathematical concepts arise from noting common activities of mathematicians, their interests and results over a given period; these in turn are presented as special cases of more general patterns.

We shall try to describe a few such new theories as illustrations of present-day research. *Information theory,* as shaped by Shannon and his successors, forms an elegant and coherent part of mathematics. In this connection we have discussed a finite set of events and the probabilities attached to them. Should the space of events be infinite (say countably infinite, or continuous) it is interesting that one still can define the notions of information theory by generalizing the measures in the space of events either by limiting processes or by introducing suitable integrations (recall the earlier discussion of measure theory). Beyond measure it has been possible to define other properties of sets situated in a space of events. If the space of events had a distance defined between its elements then, in addition to measure, one could define another property, the entropy or capacity, of sets in the space. These definitions, arising from practical problems in transmitting and coding messages, enabled mathematicians to develop general abstract theories through which some old problems in set theory have been solved. Soviet mathematicians, in particular, have had great success in using Shannon's ideas for the solution of problems in pure mathematics. It is curious that it has taken so long to generalize the idea of entropy from its original meaning in assemblies of molecules or atoms to very general combinatorial classes of events. Beyond its use in problems of communication and organization, this generalization has proved extremely valuable in abstract mathematical questions that *prima facie* seemed to have nothing to do with notions of probability!

We have mentioned the celebrated collection of problems proposed by Hilbert. One of these concerned the expression of the roots of a polynomial equation of degree n as functions of the coefficients. It is known that these roots exist as a unique set and, therefore, that they are functions of the coefficients. For equations of first, second, third, and fourth order, these functions can be expressed in a very special form: they can be obtained by applications of sums, products, and of extraction of radicals. At any rate, these functions of n variables ($n = 1, 2, 3, 4$) are representable as superpositions of functions of a smaller number of variables; at most two. The sum of any number of terms or the product of any number of terms is obtainable by repeatedly taking a sum or product of just two terms. The kth root is, of course, a function of a single variable. For $n \geqslant 5$, the roots of an algebraic equation cannot be obtained by radicals and rational operations, a famous result of Galois. Hilbert's problem was: If sums, products, and extraction of radicals are not sufficient to express the roots of equations as functions of coefficients for $n \geqslant 5$, are there other functions of a smaller number of variables such that by their repeated superposition one may obtain these roots, thus "solving" the equations? More generally, can continuous functions of many variables always be represented by superimposing functions of a smaller number of variables; say, only two? Although Hilbert was inclined to believe that this cannot be done, Kolmogorov and Arnold have proved that all continuous functions of any finite number of real variables can be represented by the superposition of continuous functions of at most two variables. These results, dating from 1957, suggested still more precise formulations and more general problems on superposition. The problems concerned analogous representation possibilities for *n-tuples* of continuous functions of n variables. For the nth-dimensional space, can one represent one-to-one continuous transformations by superposition of continuous transformations in a smaller number of variables? Under what conditions is this possible? Instead of postulating mere continuity, one may require stronger properties like repeated differentiability of functions or transformations, or their analyticity, and so on.

We have seen how useful transformations can be in formulating *qualitative* properties of motions of physical systems. A dynamical system of n mass points was represented earlier by a single point in $6n$ dimensions; the change of its configuration in time was pictured as a motion of a point in this $6n$-dimensional space. The totality of all possible initial positions, changing in time, thus represented a flow in this $6n$-dimensional "phase" space. "In general," such a flow that is volume- or measure-preserving is ergodic; that is, the representative points of the system travel with uniform density through all the available space. To explain the "in general": All such continuous measure-preserving transformations can be considered as a space of "points," each point being a transformation with a distance defined between points. The set of those transformations that are ergodic forms a "big" set, in that the set of non-ergodic transformations is representable as a sum of countably many sets that are nowhere dense in the whole space. The idea of function spaces proves useful here; it was introduced in purely abstract contexts but in many cases it serves to make precise statements about physical systems. The result on the prevalence of the ergodic property among all possible continuous, volume-preserving flows may be considered analogous to a statement that "most" real numbers are irrational or even transcendental. Of course, a specific number defined by some equations or an algorithm need not belong to this "big" set. Whether a given dynamical system possesses the ergodic property is not, in general, easily settled; but the motion of a dynamical system leads to a very *special* continuous and volume-preserving flow. J. Moser in this country and Arnold and Kolmogorov in the U.S.S.R. have found classes of dynamical systems in which (if we consider the system as a single point in the phase space) the motions do *not* pervade the whole available space, describing quasiperiodic trajectories confined to curious parts of the whole space. In other words, such physical systems have properties that fall between those of simple periodic motion (like those of a two-body Keplerian system) and those of a "general" continuous flow. If a system is "sufficiently complicated" it was believed that its motion would tend to be ergodic; roughly speaking, after enough time it would be close to any possible configuration. Some special systems need not have this property, however. For example, if the motion can be described by *linear* equations, as in vibrations of a physical system, then (at least for small amplitudes) we have periodic oscillations. Thus, an idealized elastic string will forever oscillate periodically in its initial modes. Of course, the linearity of the equations represents only an approximation to the physical situation; empirically, the elastic force is not rigorously proportional to the displacements. This force, a function of the amplitude of the displaced string, has some small terms, perhaps quadratic or of higher order, in addition to the main term to which it is proportional. If one takes this into account and follows the motions for a very long time, it might seem that the original shape of the vibration will slowly become more and more complicated. Calculation on electronic computers simulated the motions sufficiently far in time. Quite unexpectedly it was found that the vibration would not become extremely complicated, but that the string, while not exactly periodic, was still confined to a small portion of the available class of configurations. In other words, this system did not indicate general ergodicity. The results of this calculation have stimulated considerable work on such "nonlinear" problems, some of purely mathematical interest. As we have seen, a great deal is known about linear transformations of the Euclidean space. In contrast, for transformations of such a space into itself that are not linear (*e.g.*, quadratic) very little is known. Efforts in these directions may lead to the outlines of a general theory.

Game theory (introduced earlier) concerns mathematical studies of a special combinatorial character. Imagine two persons selecting, in turn, "moves" out of a given set of prescribed alternatives. After a number of these moves, the resulting configuration presents a "win" for one of the players. Game theory considers the selection of strategies when each player has to base his decisions upon the probabilities of his opponent's decisions. The first problem is to optimize the tactics of each player. Much of the theory refers to games that do not provide "complete information," those in which chance plays an essential role. Poker, as opposed to chess, would be an example of such a game. Games among more than two persons have been studied; one important problem concerns the formation of *coalitions* of players against other groups. This work originated in its modern form in a paper by J. von Neumann and was developed in a book, *Theory of Games and Economic Behavior,* by Von Neumann and Morgenstern that has initiated a whole series of mathematical researches.

A classical example of the "clairvoyance" of mathematical imagination in anticipation of developments in physical theories is the theory of Riemannian geometries. In his famous paper *Über die Hypothesen die der Geometrie Zugrunde Liegen* ["On the hypotheses which lie at the foundation of geometry"], Riemann defined and developed the theory of a whole class of geometries, generalizing "flat" Euclidean space to curved manifolds, the curvature given locally in each point of space perhaps being determined by a physical quantity like the density of matter. Quite prophetically, this proved to be what Einstein assumed many years later as a basis of his theory of general relativity. The apparatus of differential geometry, developed by Riemann and the mathematicians who followed him, was essential for the formulation of the principles of the general relativity theory. These theories are essentially "local" in character, primarily concerning the behaviour of the curvature and the geodesic lines (*i.e.,* the lines of shortest connection between points) in the neighbourhood of each point. The *global* properties of such spaces concern the topological characteristics of the space as a whole; for example, the number of k-dimensional "holes" in it (the Betti numbers). These numbers and the homotopy of these spaces (*i.e.,* the number of independent curves or surfaces lying in the space that are not contractible to each other) are topological properties of a space in its entirety. Much important research is developing the theory of such properties; differential geometry "in the large" is one of the most active parts of current mathematics. The methods used are algebraic in character and permit one to establish properties of continuous vector fields defined on such spaces.

In the same spirit of global investigations, mathematicians have studied intensively the structure of continuous groups. For example, groups of rotations of n-dimensional spaces can be considered themselves as spaces, the distance between any two rotations defined in a simple way. Recent work has succeeded in clarifying many of the topological characteristics of such groups, especially in the case where the group operation, in addition to being continuous, possesses differentiability properties. These are called Lie groups, and one tries to represent them by groups of linear transformations of the n-space; *i.e.,* to find groups of linear transformations isomorphic to the given groups. Again, the mathematical properties of such representations are found to possess important physical interpretations in quantum theory and in the theory of elementary particles. Application of these ideas to the classification of atomic spectra and of "elementary" particles again shows the unexpected powers of mathematical foresight.

The prospects that lie ahead in mathematical research (with its growing mass of individual and special results, increasing number of different "unification" attempts, and pervading mathematization of sciences and of technology) pose serious problems for there appear tendencies to erect a Tower of Babel with countless separate languages.

It is, however, helpful to consider historically the role mathematics has played in the development of other sciences. It is significant that the invention of the algorithms of differentiation and integration was contemporaneous with Newton's discovery of the laws of motion and the foundation of the dynamics of celestial bodies. All at once the fundamental laws of mechanics were formulated and the tools for drawing the consequences of these laws were invented and perfected. It can be asserted that there still is no conceptually better or technically more efficient means for formulating these laws or for calculating the motions of bodies. Differential and integral operators still form the basis of mathematical analysis. The laws of classical physics are stated in the form of differential equations, originally ordinary or total differential equations and systems of such equations. One deals with functions of a single variable; the equations relate the derivatives of the unknown functions to their values and to some given functions. The behaviour of physical systems, thus described, is predicated on "mass points," idealizations that replace mass distributions.

The mathematical treatment of continuous distributions of mass or of "fields" in physics requires *partial differential equations*. Here functions of several variables are introduced and the equations connect the partial derivatives of these functions with respect to the space variables and to time. In the 18th century the successors of Newton already dealt with such equations. These functions might be velocities in a fluid, density of matter in space, elastic stresses in a material, or temperature as it changes in space and time. Problems in hydrodynamics, elasticity, and the theory of heat could be stated and solved by partial differential equations, and throughout the 19th century mathematical physics witnessed a series of great successes in exploiting the mathematics of analysis. Later, the theory of electricity culminated in the mathematical formulation of electromagnetic phenomena by Maxwell who expressed the basic laws of electromagnetism in the form of a system of partial differential equations.

In the 19th century introduction of functions of a complex variable to physical theories brought with it almost miraculously effective algorithms to solve otherwise intractable problems. It also appeared to discover (somewhat mysteriously) a new meaning and new formulations of physical laws. Remember that the introduction of the complex variable (and of functions with complex values of an argument) came from algebra rather than from problems of the natural sciences. The calculus of variations, whose beginnings date to the 18th century, is another field of mathematical analysis. Physical laws can be formulated as postulates that certain integrals of functions of one or more variables are *extremal*. The principles of Fermat, Maupertuis, and Hamilton of shortest time for light rays or of extremal action are concise, elegant, and mathematically powerful. The theory of integral equations, examples of which already had been studied by Abel, came at the end of the 19th century. All this work forms part of mathematical analysis.

Other fields of mathematics have entered physical theories. The theory of probability forms the basis of *statistical mechanics,* which deals with the behaviour of matter; not through the mathematics of the continuum but through a model of a very great number of atoms that form a discrete assembly of interacting particles. This "combinatorial" approach parallels and supplements the continuous treatment of thermodynamics. Again in the 19th century the development of new theories of geometry prepared for the physical theories of relativity.

The new fields of mathematics have been applied to other theories in the 20th century, including Einstein's great creation of the theory of relativity. Conceptually and mathematically, the preparations for relativity already had been made. Lorentz and Poincaré considered a group of transformations of four-dimensional time-space that leave invariant the form of Maxwell's equations for electromagnetic phenomena. Einstein elevated the invariance of all equations governing physical phenomena under these transformations to a fundamental principle of physics and brought about a revolution in the concept of space and time. Astounding conclusions like that of the equivalence of mass and energy were obtained as *mathematical* consequences of this assumption. It should be stressed that $E = mc^2$ is a mathematical consequence of the invariance of laws of nature under Lorentz transformations. The freedom of construction of new mathematical schemata for theoretical physics was certainly facilitated, if not directly stimulated, by what mathematicians had done in defining and working with abstract ideas.

The work of Riemann and others, *e.g.,* on geometries more general than Euclid's, had prepared the ground and established the mathematical tools for formulating the general theory of relativity.

In quantum theory, the point of view one takes in dealing with phenomena is perhaps still more abstract. The fundamental entities or objects that form the "primitive notions" of the theory no longer are material "points" of the Euclidean space but are rather distribution functions related to "wave packets." These serve as the substratum of physical existence, and our physical observations are accessible to measurement as integrals of or operators on such distributions. Mathematically the theory is based on function spaces like the Hilbert space mentioned in Chapter 2. Long before quantum theory, Hilbert used the word "spectrum" for mathematically defined characteristics of linear transformations on his space of infinitely many dimensions. These numbers exactly correspond to the spectrum of radiation emitted by atoms!

The seeming chaos of spectral lines can be understood and ordered on the basis of a mathematical theory of groups. In mathematics itself, the idea that formal properties of groups of transformations determine and classify the objects on which they operate was applied to geometry. The famous Erlangen Program of F. Klein promulgates the tenet that a geometry is given in its essence by the group of transformations under which its objects and relations remain invariant. There now appear tendencies to generalize this approach. Much of the present work in theoretical physics continues this idea. Attempts are made to derive laws of physics and to classify "elementary particles" by means of groups of transformations. Principles of the conservation of momentum and energy, as well as conservation of charge and quantities like spin and "strangeness," are investigated by considering abstract groups.

What mathematical theories are most likely to play important roles in physical theories in the future? Phenomena on very small scale, in subatomic and nuclear dimensions, are very strange to the ideas of classical physics. Even qualitatively they require mathematical variables of a different type from the familiar real numbers and Euclidean continua. Measures cannot be obtained simultaneously and with arbitrary precision for the momentum *and* position of a particle, or for the energy *and* the time of emission of radiation. As we go to higher and higher energies in probing the constituents of matter, one may conceive of mathematical models other than those currently used in physics. Set theory (in general, point-set topology) may prove useful for constructing models of matter and radiation. Even more radically, the models by which we represent space-time itself may utilize such ideas. In astronomy, and cosmology which deals with the universe in the large, recent work has shown unexpected phenomena. Should a model of an actually infinite universe prove more adequate, the mathematical ideas of set theory will come more importantly into play. Recent discoveries in mathematical logic of the incompleteness of any axiomatic system will pose strictly scientific as well as philosophical problems regarding the universe. We have seen how in mathematics itself it has been proved in recent years that some of the fundamental problems are undecidable on the basis of existing systems of axioms. It may be that no finite system of axioms ever will be considered as definite or ultimate. Beyond purely conceptual constructions (which include mathematics), that which we call the physical world poses similar dilemmas. Should the universe actually contain an infinity of distinct "points" (stars, elementary particles of matter, or photons), statements or propositions concerning such assemblies certainly will exist that are undecidable in terms of any finite number of laws and rules stated in advance. Indeed, there are profound implications in this very recent mathematical work.

We have discussed the vital connections between mathematics and mathematical physics. Neither of these great domains of human thinking would exist in any way resembling its present form without the influence of the other. The very essence of theoretical physics lies in its mathematical formulations; the development of large parts of mathematics was stimulated and determined by problems posed by the behaviour of matter. Our ideas of space and of time are abstracted from empirical experiences. How these experiences can be amenable to mathematical treatment that can be pursued consistently and still lead to conclusions in accord with observations, may be philosophically puzzling. One reason for it is the necessary condition that measurements, and thus much of the discussion in physics or astronomy, can be reduced to operations with numbers.

Sciences that deal with the much greater variety and diversity of living forms are at present in a quite different position. We are witnessing such a rapidly increasing knowledge of primary or "elementary" biological phenomena that these are becoming ripe for mathematization. In the variety of incidental technical problems concerning the life sciences, mathematics has been a very useful tool for a long time. In problems of statistical behaviour like chemical reactions in living matter, or regularities in the behaviour of large groups of living components or of their organizations, simple calculus, algebra, and combinatorics have had many useful applications. Studies like Volterra's on changes in the numbers of individuals of living species that feed on each other were mathematically interesting beyond their implications in biology. Volterra used a system of related total differential equations that are not linear. Such work has stimulated the study of nonlinear problems in pure mathematics, where very important results continue to be obtained. Mendel's genetic laws gave rise to a number of combinatorial studies. A fair amount of mathematics is necessary to describe the behaviour of mixtures in biochemistry and, to consider the thermodynamic and quantum-theoretical bases of such processes, the great apparatus of mathematical physics is not only useful but necessary.

Beyond this, we maintain that developments in the last decade or so of biology open much larger, more intriguing, and conceptually even more promising mathematical vistas.

Important beginnings have recently been achieved in understanding operating schemas of the living cell. The generally accepted geometric DNA model of Crick and Watson is a long, helical chain of four types of molecule with linkages across the chain. In the mechanism of reproduction this "ladder" is understood to split, producing two separate chains. The molecules in each then find their complements in the surrounding material and two DNA chains are formed from the original. The arrangement of the four molecules is understood to code genetic information for the cell and the organism. The code in part represents plans for the manufacture of proteins. It also seems to contain instructions (logically speaking, of a higher order) involving *functional* behaviour, probably something of the nature of a general "flow diagram" in computing machines. Other molecules present in the cell seem to take instructions from the DNA, transferring this information to other locations where biochemical syntheses are taking place.

The exact mechanics, logic, and combinatorics of this process are not yet fully understood. New logical schemas that are established and analyzed mathematically doubtless will be found to involve patterns somewhat different from those now used in the formal apparatus of mathematics.